Marshall Cavendish Corporation
99 White Plains Road
Tarrytown, New York 10591–9001

Website: www.marshallcavendish.com

Library of Congress Cataloging-in-Publication Data

Burton, Maurice, 1898-
International wildlife encyclopedia / [Maurice Burton, Robert Burton] .-- 3rd ed.
p. cm.
Includes bibliographical references (p.).
Contents: v. 1. Aardvark - barnacle goose -- v. 2. Barn owl - brow-antlered deer -- v. 3. Brown bear - cheetah -- v. 4. Chickaree - crabs -- v. 5. Crab spider - ducks and geese -- v. 6. Dugong - flounder -- v. 7. Flowerpecker - golden mole -- v. 8. Golden oriole - hartebeest -- v. 9. Harvesting ant - jackal -- v. 10. Jackdaw - lemur -- v. 11. Leopard - marten -- v. 12. Martial eagle - needlefish -- v. 13. Newt - paradise fish -- v. 14. Paradoxical frog - poorwill -- v. 15. Porbeagle - rice rat -- v. 16. Rifleman - sea slug -- v. 17. Sea snake - sole -- v. 18. Solenodon - swan -- v. 19. Sweetfish - tree snake -- v. 20. Tree squirrel - water spider -- v. 21. Water vole - zorille -- v. 22. Index volume.
ISBN 0-7614-7266-5 (set) -- ISBN 0-7614-7267-3 (v. 1) -- ISBN 0-7614-7268-1 (v. 2) -- ISBN 0-7614-7269-X (v. 3) -- ISBN 0-7614-7270-3 (v. 4) -- ISBN 0-7614-7271-1 (v. 5) -- ISBN 0-7614-7272-X (v. 6) -- ISBN 0-7614-7273-8 (v. 7) -- ISBN 0-7614-7274-6 (v. 8) -- ISBN 0-7614-7275-4 (v. 9) -- ISBN 0-7614-7276-2 (v. 10) -- ISBN 0-7614-7277-0 (v. 11) -- ISBN 0-7614-7278-9 (v. 12) -- ISBN 0-7614-7279-7 (v. 13) -- ISBN 0-7614-7280-0 (v. 14) -- ISBN 0-7614-7281-9 (v. 15) -- ISBN 0-7614-7282-7 (v. 16) -- ISBN 0-7614-7283-5 (v. 17) -- ISBN 0-7614-7284-3 (v. 18) -- ISBN 0-7614-7285-1 (v. 19) -- ISBN 0-7614-7286-X (v. 20) -- ISBN 0-7614-7287-8 (v. 21) -- ISBN 0-7614-7288-6 (v. 22)
1. Zoology -- Dictionaries. I. Burton, Robert, 1941- . II. Title.

QL9 .B796 2002
590'.3--dc21

2001017458

Printed in Malaysia
Bound in the United States of America

07 06 05 04 03 02 01 8 7 6 5 4 3 2 1

Brown Partworks
Project editor: Ben Hoare
Associate editors: Lesley Campbell-Wright, Rob Dimery, Robert Houston, Jane Lanigan, Sally McFall, Chris Marshall, Paul Thompson, Matthew D. S. Turner
Managing editor: Tim Cooke
Designer: Paul Griffin
Picture researchers: Brenda Clynch, Becky Cox
Illustrators: Ian Lycett, Catherine Ward
Indexer: Kay Ollerenshaw

Marshall Cavendish Corporation
Editorial director: Paul Bernabeo

Authors and Consultants

Dr. Roger Avery, BSc, PhD (University of Bristol)

Rob Cave, BA (University of Plymouth)

Fergus Collins, BA (University of Liverpool)

Dr. Julia J. Day, BSc (University of Bristol), PhD (University of London)

Tom Day, BA, MA (University of Cambridge), MSc (University of Southampton)

Bridget Giles, BA (University of London)

Leon Gray, BSc (University of London)

Tim Harris, BSc (University of Reading)

Richard Hoey, BSc, MPhil (University of Manchester), MSc (University of London)

Dr. Terry J. Holt, BSc, PhD (University of Liverpool)

Dr. Robert D. Houston, BA, MA (University of Oxford), PhD (University of Bristol)

Steve Hurley, BSc (University of London), MRes (University of York)

Tom Jackson, BSc (University of Bristol)

E. Vicky Jenkins, BSc (University of Edinburgh), MSc (University of Aberdeen)

Dr. Jamie McDonald, BSc (University of York), PhD (University of Birmingham)

Dr. Robbie A. McDonald, BSc (University of St. Andrews), PhD (University of Bristol)

Dr. James W. R. Martin, BSc (University of Leeds), PhD (University of Bristol)

Dr. Tabetha Newman, BSc, PhD (University of Bristol)

Dr. J. Pimenta, BSc (University of London), PhD (University of Bristol)

Dr. Kieren Pitts, BSc, MSc (University of Exeter), PhD (University of Bristol)

Dr. Stephen J. Rossiter, BSc (University of Sussex), PhD (University of Bristol)

Dr. Sugoto Roy, PhD (University of Bristol)

Dr. Adrian Seymour, BSc, PhD (University of Bristol)

Dr. Salma H. A. Shalla, BSc, MSc, PhD (Suez Canal University, Egypt)

Dr. S. Stefanni, PhD (University of Bristol)

Steve Swaby, BA (University of Exeter)

Matthew D. S. Turner, BA (University of Loughborough), FZSL (Fellow of the Zoological Society of London)

Alastair Ward, BSc (University of Glasgow), MRes (University of York)

Dr. Michael J. Weedon, BSc, MSc, PhD (University of Bristol)

Alwyne Wheeler, former Head of the Fish Section, Natural History Museum, London

Picture Credits
Heather Angel: 1342, 1404, 1405, 1407; **Ardea London:** P. Morris 1325, 1328, 1377, M.D. England 1382, Jean-Paul Ferrero 1337, 1387, 1388, Francois Gother 1353, 1354; **Art Explosion:** 1398; **Neil Bowman:** 1345, 1369, 1416; **Bruce Coleman:** Theo Allofs 1374, Ingo Arndt 1411, 1412, Ken Balcomb 1350, George Bingham 1386, Nigel Blake 1415, Mr. J. Brackenbury 1392, Fred Bruemmer 1384, Jane Burton 1302, 1326, 1330, 1365, Bob and Clara Calhoun 1339, 1340, John Cancalosi 1305, 1309, 1336, 1344, 1361, 1378, 1379, 1385, Bruce Coleman Inc 1397, Alain Compost 1383, Sarah Cook 1391, 1402, 1434, Gerald S. Cubitt 1343, 1437, Dr. P. Evans 1401, Paolo Fioratti 1357, M.P.L. Fogden 1341, 1362, 1363, Jeff Foott 1417, C.B. and D.W. Frith 1429, Peter A. Hinchliffe 1420, Charles and Sandra Hood 1327, 1322, HPH Photography 1389, Johnny Johnson 1352, 1358, 1360, Rob Jordan 1371, Janos Jurka 1375, Steven C. Kaufman 1418, P. Kaya 1394, Wayne Lankinen 1304, Werner Layer 1333, 1410, Robert Maier 1368, Luiz Claudio Marigo 1312, 1364, 1356, George McCarthy 1348, Joe McDonald 1435, 1346, Michael McKavett 1319, Rita Meyer 1355, Pacific Stock 1321, 1351, Allan G. Potts 1301, 1308, 1347, 1433, Hans Reinhard 1334, 1359, 1366, 1400, Jens Rydell 1317, Dr. Frieder Sauer 1320, John Shaw 1380, Kim Taylor 1393, 1395, 1396, 1419, Norman Tomalin 1335, Jan Van de Kam 1307, Dries Van Zyl 1390, Colin Varndell 1300, 1318, Staffan Widstrand 1310, 1311, Rod Williams 1323, 1329, 1436; **Chris Gomersall:** 1367, 1370; **Ben Hoare:** 1399; **Natural Visions:** Geoff Moon 1373, Slim Sreedharan 1315; **NHPA:** A.N.T. 1372, Agence Nature 1403, Anthony Bannister 1427, G.I. Bernard 1313, Stephen Dalton 1428, Martin Harvey 1425, 1426, Daniel Heuclin 1324, Peter Parks 1406, Rod Planck 1306, Kevin Schafer 1381, John Shaw 1332, Norbert Wu 1408; **Planet Earth Pictures:** Peter David 1414, Carol Farneti 1314, Adam Jones 1349, Brian Kenney 1331, Ken Lucas 1376, 1421, 1422, 1430, 1431, Doug Perrine 1424, Norbert Wu 1413; **Still Pictures:** W. Fischer 1423, Klein/Hubert 1432. **Artwork:** Catherine Ward 1409.

Contents

JACK DEMPSEY

THIS VISUALLY DAZZLING FISH once enjoyed, among aquarists, a high popularity that persists today, albeit to a lesser degree. The Jack Dempsey is aggressive and can create havoc in a mixed tank with other kinds of fish, so it needs to be kept in a separate aquarium. The fighting between males is ritualized and resembles a boxing match. The species was named after the world heavyweight boxing champion of 1919–1926, when it first gained popularity.

Young fish are brown or gray with several transverse bands. Mature males, which are up to 10 inches (25 cm) long, are deep brown to black, with a bluish sheen on the flanks and a scattering of silvery or green dots. They have a bold black blotch, usually ringed with gold, at the edge of the gill cover, and another at the base of the tail. The dorsal fin is yellow-bordered. Females are slightly smaller than males; usually duller in color, they have fewer and fainter spots, as well as shorter fins. In both sexes the body is laterally flattened (relatively narrow for its height). The large head has a jutting lower jaw. Mature males have a bulging forehead.

Bully boy

The Jack Dempsey lives in sluggish freshwater habitats in the American tropics and subtropics. Little is known about its lifestyle in the wild, but captive study has revealed much about its strategies of fighting and breeding, two aspects of its behavior that are closely linked. When a male comes into breeding condition, he establishes a territory. Should another male swim into that territory, the owner swims toward him to begin what is known as a lateral display: swimming close beside him, nose to tail. At the same time, the owner raises his dorsal and anal fins, spreads his paired fins and raises his gill covers. From the side he now looks much larger. Meanwhile, his colors brighten, masking the two black spots. The overall effect is to present his opponent with a terrifying and daunting spectacle.

The intruder may retreat, in which event he is chased out of the territory. More usually, he responds by raising his own fins, and his own colors grow brighter. In that event the two circle like boxers, each trying to jab the other with the sharp edge of the jaw. The two may later seize each other by the mouth in a trial of strength. In the end it is normally the intruder that yields, being then chased out by the resident. Injuries sustained in the fight are usually slight.

Feminine pacifism

Should a female wander into the territory, something of the same sort takes place, but everything depends on how near she is to being ready to spawn. In any case, the male displays at her as if she were a male. Instead of raising her fins, however, she lowers them in what is called a show of symbolic inferiority (a tactic practiced also by immature intruders). This does not prevent him from attacking her, butting her with his jaw, but she accepts the blows without retaliating. If she is not ready to spawn, she will eventually be chased away. But if she is ready, the male becomes less hostile and accepts her as a mate.

In the open waters of her wild habitat an unready female would have room to escape a territorial male. In an aquarium, she is treated as any other invader and may be wounded. The usual procedure in bringing a male and a female together in a tank is to insert a pane of glass to separate them and so prevent physical contact.

The two fish display to one another through the glass. In due course, and as the female comes into breeding condition, their confrontation turns to courtship. When the glass is finally taken out, they come together peacefully as a pair to spawn.

Jack Dempseys of both sexes are by nature aggressive. Females are particularly fierce in the defense of young, while males (below) use violent tactics to run intruders off their personal territory.

JACK DEMPSEY

CLASS **Osteichthyes**

ORDER **Perciformes**

FAMILY **Cichlidae**

GENUS AND SPECIES ***Cichlasoma octofasciatum***

LENGTH
Up to 10 inches (25 cm)

DISTINCTIVE FEATURES
Deep, elongated body; large head; jutting lower jaw. Male: dark brown or black with bluish sheen on flanks; bold black spot by gill cover and another at base of tail; silver or green spots on head, flanks and tail; fins dark, with yellow border on dorsal fin. Female: duller in color; fainter spots. Young: brown or gray with transverse bars.

DIET
Algae, insects, mollusks, crayfish and some other fish

BREEDING
Age at first breeding: when around 3 in. (7.5 cm) in length; number of eggs: 500 to 800; hatching period: about 2 days; breeding interval: not known

LIFE SPAN
Not known

HABITAT
Sandy or muddy beds of sluggish rivers, swamps and drainage ditches

DISTRIBUTION
Native range: American Tropics, from northern Mexico south to Rio Negro and Amazon River Basin in Brazil. Introduced range: now widespread in Florida; smaller colonies in California and Connecticut.

STATUS
Locally common across much of range

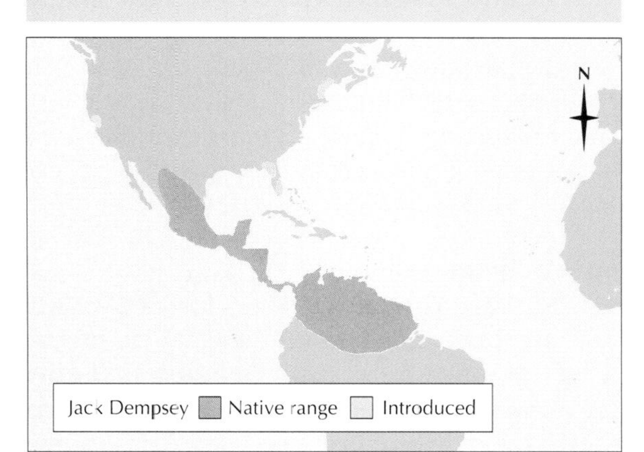

Working off energy

Coming into breeding condition means more than merely getting ready to spawn. In the male in particular there is a buildup of stored energy, much of which is dissipated in fighting. In many species of cichlid fish the male uses his fins to dig pits in the sand, as described for the firemouth, *Cichlasoma meeki*. Studies of Jack Dempseys in captivity indicate that one adult, more usually the female, digs a single large pit.

Eventually both male and female Jack Dempseys choose a flat surface and start to clean it with their mouths. The female then moves over the area and lays her eggs, which stick to the cleaned surface. The male follows her and fertilizes the eggs. The eggs take 50 hours to hatch, during which time both parents fan them with their fins. For some 96 hours after hatching, the babies, unable to swim, feed on their yolk sacs. At this stage they are known as wrigglers. Each parent takes one wriggler at a time in its mouth and places the tiny charge in the pit in the sand. Here the parents continue to guard their family. As the babies begin to swim out of the pit, the parents pick them up in the mouth and spit them gently back into the mass.

There comes a time, however, when the young swimmers are too big for the parents to keep spitting them back, so they give up doing so. Instead, they direct their efforts to keeping their family of several hundred bunched together for protection. Some 500 to 800 or more eggs may be laid in a season.

Family cannibalism

Scientists have attempted to find out whether Jack Dempseys recognize their own babies. They have taken away a pair's eggs and replaced them with eggs laid by a related species. From these experiments it seems that, providing the foster broods grow at the same rate and are about the same size, all will be well. The Jack Dempseys appear not to recognize their own babies but instead identify the family they are guarding by its uniform size and growth rate.

Sometimes, in swapping the clutches of eggs, the scientists have overlooked the odd batch, particularly any that are secreted beneath slabs of rock. On these occasions, when hatching time arrives the wrigglers are a mixed brood of Jack Dempseys and babies of another species. All goes well at first. After a few weeks, however, with one set growing a little more rapidly than the other, the two species can be distinguished. The larger young become aware that their smaller broodmates can be overpowered, and despite the fact that they have lived together peaceably until now, the more powerful babies eat their foster brethren.

JACK RABBIT

The white-tailed jack rabbit is abundant on the great plains and prairies of North America. It is a game animal in many areas.

The jack rabbits of North America are in fact hares belonging to the genus *Lepus*, close relatives of the brown or common hare, *L. europaeus*, and snowshoe hare, *L. americanus*. The white-tailed jack rabbit, *L. townsendii*, also known as the plains or prairie hare, has a brownish coat in the summer, which changes to white in the winter. Only the 6-inch (15-cm) black-tipped ears and 4-inch (10-cm) white tail remain black year-round.

The white-tailed jack rabbit, which weighs up to 10 pounds (4.5 kg), lives in the prairies of the northwest. To the south lives the smaller black-tailed jack rabbit or jackass hare, *L. californicus*. The latter name is derived from the 8-inch (20-cm) black-tipped ears. The coat is sandy except for the black upper surface of the tail. It does not turn white in winter. This species lives in the arid country from Oregon to Mexico and eastward to Texas. A small population in Florida is descended from imported jack rabbits, used in training greyhounds, that have gone wild.

The other two species of jack rabbit are the antelope jack rabbit, *L. alleni*, of southern Arizona and northwestern Mexico, and the white-sided jack rabbit, *L. callotis*, of southwestern New Mexico and parts of Mexico.

Safety in speed

Like all hares, jack rabbits do not live in burrows. The exception is the white-tailed jack rabbit, which during winter burrows under the snow for warmth and to gain protection against predators such as owls. Otherwise jack rabbits escape

JACK RABBITS

CLASS **Mammalia**

ORDER **Lagomorpha**

FAMILY **Leporidae**

GENUS AND SPECIES **Black-tailed jack rabbit, *Lepus californicus*; white-tailed jack rabbit, *L. townsendii*; antelope jack rabbit, *L. alleni*; white-sided jack rabbit, *L. callotis***

ALTERNATIVE NAMES
Black-tailed jack rabbit: jackass hare; white-tailed jack rabbit: plains hare; prairie hare

WEIGHT
3–15½ lb. (1.35–7 kg)

LENGTH
Head and body: 15¾–27½ in. (40–70 cm); tail: 1½–4 in. (3.5–10 cm)

DISTINCTIVE FEATURES
Gray brown upperparts; white or cream underparts; thick fur; furred feet; long ears; some species turn white in winter

DIET
Grasses, herbs, buds, twigs, bark; carrion

BREEDING
Age at first breeding: 2 years; breeding season: variable; number of young: 1 to 6; gestation period: 41–48 days; breeding interval: several litters per year

LIFE SPAN
Up to 5 years

HABITAT
Grassy areas, scrub, farmland, desert and forest margins, tundra

DISTRIBUTION
North America, from southern Canada to Mexico, excluding eastern U.S. seaboard

STATUS
White-sided jack rabbit: near threatened; other species common

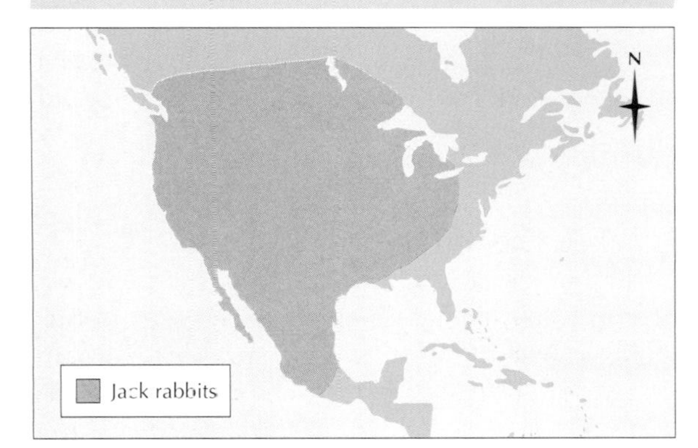

The black-tailed jack rabbit's large ears act as radiators. A network of veins close to the skin allows excess body heat to escape.

detection by crouching among the sparse vegetation of the prairies and semidesert terrain. They stay out of sight in shade during the day and emerge toward evening.

Within its home range each jack rabbit has several shallow scrapes in the ground, known as forms, which are shaded and concealed by vegetation. If flushed, a jack rabbit runs very fast, sometimes exceeding 45 miles per hour (70 km/h) in a series of low, 20-foot (6-m) bounds. Every so often it leaps 4–5 feet (1.2–1.5 m) into the air to clear vegetation and check for danger.

Water from cacti

Jack rabbits feed mainly on grass and plants such as sagebrush or snakeweed and often become serious pests where their numbers build up. To protect crops and to save the grazing for domestic stock, hunts are organized or poisoned bait is put down. In the arid parts of their range, when the grass has dried up, jack rabbits survive on mesquite and cacti. They can get all the water they need from cacti provided they do not lose too much moisture in keeping cool. To eat a prickly cactus, a jack rabbit carefully chews around a spiny area and pulls out the loosened section. Then it puts its head into the hole and eats the moist, fleshy pulp that it finds inside.

Born in the open

The length of the breeding season varies according to the latitude of the population, being shorter in the north. At the onset of breeding, jack rabbits indulge in the typical madcap antics of hares. The males chase and fight each other. They rear up, growling, and batter each other with their forepaws. They also bite each other, tearing out tufts of fur or even flesh, and deliver the occasional violent kicks with the hind legs. A well-aimed kick can badly wound the recipient; otherwise the fight continues until one of the combatants turns tail and flees. The males also chase potential mates, occasionally injuring them with kicks or bites in their eagerness to breed.

The baby jack rabbits are born in open nests concealed by brush or grass and lined with fur that the female pulls from her body. The litters are usually of three or four young, but there may be as few as one or as many as six. The newborn young weigh 2–6 ounces (56–170 g), are well furred and have open eyes. They can stand and walk a few steps immediately after birth, but they do not leave the nest for about 4 weeks.

Radiator ears

Large ear flaps are a characteristic of desert animals, such as the bat-eared fox, *Otocyon megalotis*. Not only do they improve hearing, they also cool the body. In a jack rabbit's ears, which are up to 7 inches (17 cm) long in some species, a network of veins carries blood close to the skin, allowing excess body heat to escape. This adaptation relies on the ears being warmer than the air. A difference of a few degrees is sufficient, so it is extremely effective in the semiarid home of the black-tailed jack rabbit, where a clear sky may have a temperature of 50–59° F (10–15° C), compared with 100° F (38° C) in the jack rabbit's ears. Jack rabbits rely on radiation to keep them cool because they do not get enough water to be able to use evaporation as a means of cooling. In hot weather they also make use of every bit of shade. In their forms the ground temperature is lower than both the air and body temperature.

White-tailed jack rabbits turn white or pale gray in winter, except in the far south of their range. They lie low for warmth and to hide from predators.

JAEGER

A long-tailed jaeger in flight. Jaegers are somewhat smaller than their close relatives the skuas. They also exhibit light and dark color phases, whereas skuas are always dark brown.

These relatives of the gulls and terns can be described as gulls that have turned into hawks. Several gulls are hunters, and many of them are opportunistic feeders with varied diets, but none has become so dependent on catching birds and mammals as the jaegers and their close relatives the skuas. Their bills are like those of the gulls but bear a hook. Skuas and jaegers are unique because they are the only birds to combine webbed feet, for swimming, with strong, hooked claws, for subduing prey.

Two of a kind: jaegers and skuas

There are three species in each of two genera. In North America those belonging to the genus *Stercorarius* are known as jaegers, while those in *Catharacta* are called skuas. The word "jaeger" is derived from the German for hunter. The largest species is the great skua, *Catharacta skua*, popularly known as the bonxie in Scotland. It is 1¾ to nearly 2 feet long (53–58 cm), with a wingspan of 4¼–4⅗ feet (1.3–1.4 m). It is a little longer than the herring gull, *Larus argentatus*, and more heavily built. From a distance the plumage appears uniformly dark brown, and in flight the white patches at the base of the primary wing feathers confirm its identity. Up close, the plumage is streaked with rufous brown and white, and occasionally the head and neck may be totally white. The two other species of the same genus are similar, being large, heavyweight, dark brown birds. However, the Chilean skua, *C. chilensis*, is more rufous or even cinnamon brown.

The three jaegers are smaller, ranging from 1⅓–1¾ feet (41–53 cm) long, depending on species, and are less heavily built. In these the two middle tail feathers are longer than the rest, only slightly longer in the parasitic jaeger, *Stercorarius parasiticus*, but up to one third of the total body length in the long-tailed jaeger, *S. longicaudus*. The long tail feathers of the pomarine jaeger, *S. pomarinus*, can twist through 90°. The plumage of the jaegers is also brown, but there are light and dark color phases. Individuals may be uniformly brown or may have grayish white or brownish white bodies with brown wings, tails and crowns. Dark phase long-tailed jaegers are very rare, and in the Arctic most of the parasitic jaegers are light.

The great skua breeds in the North Atlantic, where it nests in Iceland, the Faeroes and Scotland. Before 1930 most northern great skuas bred in Iceland, but there has been a rapid increase of breeders in the Shetland Islands and a spread into neighboring parts of Scotland. The South Polar skua, *C. maccormicki*, nests in the Southern Hemisphere around the fringes of the Antarctic continent, on Antarctica and subantarctic islands, in the extreme south of South America and on

A Falkland skua, ***Catharacta skua antarctica,*** *a subspecies of the great skua, feeds on a penguin chick. Skuas and jaegers are unique in being the only birds to have both webbed feet and claws.*

the Falkland Islands. The jaegers have a circumpolar distribution in the Northern Hemisphere, mainly in the tundra areas of North America and Eurasia, although the parasitic jaeger nests as far south as Scotland.

From ocean to ocean

Outside the breeding season jaegers and skuas are rarely seen on land, for they lead a pelagic life well out to sea. Most migrate toward the Tropics and can sometimes be seen flying along coasts. It has been known for some time that some jaegers and skuas cross the equator on their migration, for the South Polar skua has been seen off Japan. It also spends southern winters in the North Pacific and North Atlantic generally. The jaegers that breed in the Northern Hemisphere, meanwhile, are regularly seen in Australian waters.

Bullies of the sea

Jaegers and skuas are basically fish-eaters, and fish make up the bulk of their food when they are out at sea. They also fish during the breeding season, and in much of the Antarctic South Polar skuas can be seen returning to their nests with their feathers encrusted with ice from diving into the freezing sea.

PARASITIC JAEGER

CLASS **Aves**

ORDER **Charadriiformes**

FAMILY **Stercorariidae**

GENUS AND SPECIES ***Stercorarius parasiticus***

ALTERNATIVE NAME
Arctic skua (Britain only)

WEIGHT
¾–1¼ lb. (330–570 g)

LENGTH
Head to tail: 1⅓–1½ ft. (41–46 cm); wingspan: 3⅗–4 ft. (1.1–1.25 m)

DISTINCTIVE FEATURES
Robust body; strong, hooked bill; long, narrow wings; webbed feet with strong claws. Adult pale phase: grayish white neck and underparts; brown crown, back and wings. Adult dark phase: entirely dark brown. Juvenile: brown with pale markings.

DIET
Mainly fish (often stolen from other birds); also bird eggs, nestlings, insects and carrion

BREEDING
Age at first breeding: 4–5 years; breeding season: eggs laid May–July; number of eggs: usually 2; incubation period: 25–28 days; fledging period: 25–30 days; breeding interval: 1 year

LIFE SPAN
Up to 18 years

HABITAT
Seas and oceans, often far from land; nests on coastal tundra and moors

DISTRIBUTION
Summer: northern North America, Greenland, Iceland and northern Eurasia. Winter: Pacific and Atlantic Oceans.

STATUS
Uncommon

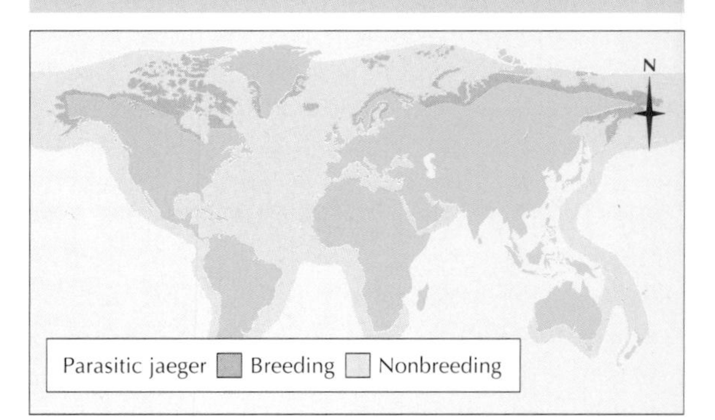

Much of their food is, however, obtained by hunting or piracy. All skuas and jaegers are famous for kleptoparasitism (literally food parasitism, or food-stealing). They have very fast, falconlike flight and simply outmaneuver and harass other birds into dropping their catch. Skuas will, for example, waylay seabirds such as cormorants and gannets. They tweak the wings or tails of gannets as they fly back with fish for their chicks, so knocking the gannets off balance. In their hurry to regain themselves, the gannets regurgitate their food, which is swiftly caught by the skuas. Great skuas also prey on birds such as puffins, kittiwakes and prions, and the jaegers prey on wheatears, buntings, swallows and small mammals. The eggs and young of many birds are also taken, as are insects and carrion.

Penguin predators

In the Antarctic, South Polar skuas are the chief predators of the eggs and chicks of penguins. Although they appear to be savage predators, the skuas are, in fact, doing a service as they mainly take the eggs and chicks that would perish anyway. They steal eggs that are not properly guarded and kill any chicks that are already weak. South Polar skuas will also combine to harass a sitting penguin, one skua luring the penguin off its nest while the other steals up behind and seizes the egg. South Polar and Chilean skuas are also known to take stillborn baby seals and afterbirths from seal nurseries.

Dive-bombers

The breeding habits of jaegers and skuas are similar to those of gulls and terns but they do not nest in such dense colonies. Each pair defends quite a large area, in the middle of which is the nest, a depression in moss or grass, scantily lined with grass, moss or roots. Both sexes incubate the two green-brown, blotched eggs, usually sitting for less than 1 hour at a time. The incubation period for the great skua is about 28 days and for the jaegers 25–28 days. Before the eggs are laid and during incubation the male feeds the female, but when the chicks hatch they are fed by both parents in the same way as gull and tern chicks are fed: the chicks peck at their parents' bills to induce them to regurgitate food.

Both eggs and chicks are vigorously defended by the parents. Skuas and jaegers are even more vigorous in defense of their nests than gulls and terns, swooping repeatedly at intruders. The great skua weighs 3–4 pounds (1.4–1.8 kg), and the effect of a pair diving almost vertically at one's head from about 50 feet (15 m), calling raucously at the same time, is most alarming. It is made more alarming by the fact that the skuas sometimes hit the intruder with their feet or wings. The jaegers are also frightening. They lack the weight of a great skua, but this is compensated by the speed of their flight. They also perform distraction displays, beating their wings on the ground and calling. Combined with the attacks and the coloring of the eggs and chicks, this makes their nests very difficult to find.

Seizing the opportunity

Unlike hawks, jaegers and skuas cannot hold their food down with their feet, so they are not as efficient at disposing of their prey. Most are basically fish-eaters that feed elsewhere when the opportunity arises. In the Arctic, however, pomarine jaegers, and to a lesser extent parasitic and long-tailed jaegers, are dependent on lemmings. In years when lemmings are scarce the jaegers may fail to breed. This represents one of the classic predator/prey cyclical population curves.

Great skuas in an aggressive display. Skuas and jaegers are known for their vigorous defense of both eggs and chicks, directly attacking intruders and carrying out distraction displays.

JAGUAR

Jaguars are strong climbers and often rest on a low branch to get a good view of their surroundings.

THE JAGUAR IS THE LARGEST of the American cats. Although no longer than the leopard, *Panthera pardus,* of Africa and Asia, it is more heavily built: the head and body measure up to 6 feet (1.9 m) long and the tail extends to about 3 feet (1 m) long. An adult may weigh up to 300 pounds (135 kg). The ground color of the coat is yellow, becoming paler underneath. All over the body is a pattern of black spots up to 1 inch (2.5 cm) in diameter. The jaguar's coat is usually easy to distinguish from that of a leopard because the spots of the former are arranged in a rosette of four or five around a central blotch. The leopard's spots, by contrast, lack a central blotch. The jaguar's rosettes are not so marked on the legs or head, where the spots are more dense.

Occasionally, melanistic (black) and albino (white) jaguars may be seen. Melanism, which occurs widely among leopards and occasionally servals, too, is caused by a recessive gene and results from an excess of the pigment melanin. In very strong sunlight the spots are just visible on a black jaguar.

Furtive forest cat

The jaguar ranges from southern Mexico to northern Argentina. Known in South America as *el tigre,* its favored habitat has much in common with the tiger. Both like dense forest, especially where there are watercourses, but come into more open country when food is abundant there. Jaguars are adaptable and can be found from the plains to the high Andean plateaus.

Jaguars are good climbers, rivaling the leopard in their ability to prowl through trees, often stalking prey along branches. In areas that are flooded for part of the year, jaguars are confined to trees—except when they take to water, for they are excellent swimmers as well.

In the dappled shade of a rain forest a jaguar's spotted coat blends in so well with its surroundings that this elusive cat is almost impossible to see, let alone study. What little is known about it is based mainly on the stories of South American Indians and the accounts of hunters. However, a conservation plan established in 1993 is now beginning to yield more concrete information regarding the species' life history.

An adult jaguar holds a territory or range that it defends against other jaguars. Marking it out by raking its claws down tree trunks and depositing pungent urine and dung, the jaguar then patrols the area regularly. The scent marks have a double function, however, serving also to inform neighboring jaguars of the resident's sexual condition. The jaguar's territoriality explains the reports of travelers in the South

JAGUAR

CLASS **Mammalia**

ORDER **Carnivora**

FAMILY **Felidae**

GENUS AND SPECIES ***Panthera onca***

ALTERNATIVE NAMES
Onca; el tigre (Latin America only)

WEIGHT
Up to 300 lb. (135 kg); male heavier than female

LENGTH
Head and body: 5–6 ft. (1.5–1.9 m); tail: 1½–2½ ft. (45–75 cm)

DISTINCTIVE FEATURES
Strong, muscular body; yellow to rufous coat with paler belly; black spots on head and limbs; rosettes on back and flanks; some individuals completely melanistic (totally black)

DIET
Mainly mammals, such as capybaras, tapirs, peccaries, sloths, monkeys and deer; also caimans, alligators, fish, turtles and eggs

BREEDING
Age at first breeding: 2–4 years; breeding season: potentially all year, but births peak in local rainy season; gestation period: 93–105 days; number of young: 1 to 4; breeding interval: 2–4 years

LIFE SPAN
Up to 24 years in captivity; half that in wild

HABITAT
Mainly riverside forest and savanna; also scrub and arid terrain

DISTRIBUTION
Mexico south to northern Argentina

STATUS
Near threatened; declining in many areas

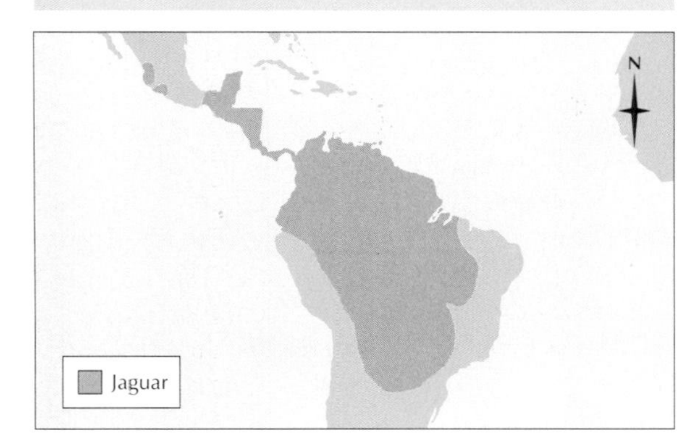

The jaguar looks similar to the leopard, but in terms of its habitat preferences and hunting tactics, it actually has more in common with the tiger.

American forests, who describe being tailed by a jaguar for many miles and then suddenly abandoned. It is thought that, in each case, the jaguar had been seeing the traveler off its territory.

Top predator

At the peak of the food chain wherever it lives, the jaguar has the pick of a wide variety of food. Its main prey includes capybaras and peccaries, but it also tackles large mammals such as tapirs and domestic cattle as well as sloths, anteaters, monkeys and deer. A jaguar may even kill and eat an alligator; and when a freshwater turtle comes ashore to lay its eggs, the big cat tips it over and rips into its belly, also devouring any of the eggs that it can dig from the sandy riverbank.

Another favorite is fish. The jaguar lies in wait on a rock or a low branch overhanging the water, then scoops out a passing victim with a deft paw. Locals claim that the jaguar has been known to dip its tail in the water, using it as a lure. Although this sounds unlikely, it is possible that a jaguar trailing its twitching tail-tip in the water may unwittingly attract an inquisitive fish. After one or two chance successes, the jaguar may perhaps learn to perform the trick deliberately.

It is thought that jaguars "take possession" of their prey and defend it. Indians in Guyana have told of jaguars attaching themselves to herds of peccaries and following them about, preying on those that become separated from their fellows.

Year-round breeding

Jaguars breed at any time of the year, although there are seasonal peaks depending on locality. Females have cubs no more than every other year. Gestation lasts about 100 days, and the mother bears one to four cubs in deep cover. Born sightless, the cubs open their eyes at 2 weeks and suckle for up to 6 months. Although they enjoy the mother's protection for up to a year and a half, many (up to 50 percent of large litters) die before reaching adolescence. Sexually mature at 2–4 years, the young are driven away by their mother to find territories of their own. Jaguars have lived as long as 24 years in captivity.

Conflict with humans

There are inevitably a few reports of man-eating jaguars. Some claim that the jaguar is more dangerous than the lion or leopard, partly because of its habit of defending a territory and partly because it becomes possessive over kills. When the prey is livestock, nearby humans are at risk. Not surprisingly, jaguars have been said to defend themselves viciously against hunters and their dogs. In exoneration of the jaguar, the explorer and naturalist Alexander von Humboldt tells of two Indian children playing in a forest clearing. They were joined by a jaguar, which bounded around them until it accidentally scratched the face of one child. The other child seized a stick and struck the jaguar on the face; the wild beast merely slunk back into the jungle. How much truth lies in the story is impossible to say, but it is not the first account of a large carnivore playing with children.

In reality it is the jaguar which has most to fear from humans. For centuries it has been shot for its lustrous spotted coat, and although the fur trade has been subjected to increasingly stringent controls by international legislation there are still an unknown number taken illegally each year.

A more insidious threat comes in the form of habitat loss. Deforestation for timber is a major threat in Latin America, and protective legislation is hard to enforce. Cattle ranchers also kill the occasional jaguar, seeing in the cat a threat to their cattle and horses, both of which it is capable of killing and dragging off into cover. Although some jaguar populations, such as those in Belize, are healthy, the species' overall range has dwindled alarmingly over the last half-century. It is now absent from most of Argentina, Uruguay and eastern Brazil, as well as from former haunts in Arizona and Mexico.

The jaguar is still widespread, but its range is shrinking and it is extremely rare in many places. Only 500 jaguars now remain in Mexico.

JAGUARUNDI

THE JAGUARUNDI IS UNLIKE any other member of the cat family, Felidae, more closely resembling a large weasel in shape and habits. Its body is long, its legs are very short and its head is flattened with small ears and a sloping face. Most forest cats are spotted or striped to some degree, but the jaguarundi's short coat is a uniform rusty red or gray, depending on the habitat. The peculiar appearance of this small cat is attributed by some scientists to its descent from an ancestral form of the puma. The jaguarundi and the modern puma and jaguar each have 38 chromosome pairs; all other South American felids (cats) have 36.

A jaguarundi, sometimes spelled jaguarondi, may be up to 54 inches (1.37 m) long, including a tail of up to 24 inches (60 cm), but its shoulder height is no more than 12 inches (30 cm). An adult weighs up to 20 pounds (9 kg).

Jaguarundis are secretive residents of lowland forest, forest edges, dense brushland and scrub. They favor running water in their habitat. The species' range extends from the southern border of the United States, through the whole of Central America, to north-central Argentina in South America. There are reports of sightings in Texas, New Mexico and Florida; the last-named are probably descendants of reintroduced stock.

Otterlike cat

The jaguarundi's name is derived from the name used for this species by the Tupi, an aboriginal people of Brazil. In other parts of America it is called the otter cat, because its appearance has been likened to that of an otter or weasel. Moreover, the hunting habits of this day-active animal are less like those of a cat than a member of the mustelid family. Instead of stalking prey or lying in wait like a typical cat, it gives chase. An excellent runner despite its short legs, a jaguarundi sprints for 1 mile (1.6 km) or more if necessary, even through dense undergrowth, running down rodents, reptiles and ground birds. Its compact form enables it to pursue victims into tight corners.

A jaguarundi readily takes to water, another trait rare among cats, and can climb trees well on account of its low center of gravity.

The jaguarundi's elongated body, short legs and flattened head give it a quite uncatlike appearance. It looks more like a large otter or weasel.

The jaguarundi is unusual because it is uniformly colored. Most forest cats have camouflage markings such as spots, blotches or stripes.

JAGUARUNDI

CLASS **Mammalia**

ORDER **Carnivora**

FAMILY **Felidae**

GENUS AND SPECIES ***Felis yagouaroundi***

ALTERNATIVE NAMES
Jaguarondi; otter cat; weasel cat; gato; eyra

WEIGHT
10–20 lb. (4.5–9 kg)

LENGTH
Head and body: 21–30½ in. (53–77 cm); tail: 13–24 in. (33–60 cm)

DISTINCTIVE FEATURES
Long, slender body; flattened head with sloping face and small ears; short legs; long tail; numerous morphs and subspecies with a variety of coat colors, from dark gray to chestnut red

DIET
Small mammals, birds, reptiles, amphibians, arthropods and fish

BREEDING
Age at first breeding: 2–3 years; breeding season: variable; gestation period: 60–75 days; number of young: 1 to 4; breeding interval: 1–2 years

LIFE SPAN
Up to 15 years

HABITAT
Lowland forest and other dense cover

DISTRIBUTION
From northern Mexico or extreme southern U.S., through Central America to Argentina; absent from Pacific coast of South America

STATUS
Rare or scarce: 4 subspecies; endangered: other 4 subspecies (including *F. y. carcomitli* of Mexico and southernmost Texas)

Morphs and subspecies

Different color morphs occur among jaguarundis: individuals may be red or gray, but not both. The darker color morphs are often associated with animals living in dense forest habitat, although this is not always the case. The red color morph was once thought to be a separate species, classed *Felis eyra*. That name is now used for one of eight subspecies, *F. yagouaroundi eyra*, which is found in Brazil, Paraguay and Argentina. The others are *F. y. armeghinoi* of Argentina; *F. y. carcomitli*, Mexico (and possibly southern Texas); *F. y. fossata*, Mexico and Honduras; *F. y. melantho*, Peru and Brazil; *F. y. panamensis*, Nicaragua south to Ecuador, *F. y. tolteca*, Mexico; and *F. y. yagouaroundi*, Guiana and Amazonia.

There is no strict breeding season, and after a 10-week gestation period a female gives birth to up to four kittens. The newborns wear spotted coats to conceal them, but these soon fade. They move from milk to solids at 6 weeks of age.

The jaguarundi's plain coat has helped save it from the worst depredations of the fur trade, but like most American felids its habitat is under threat, and a number of subspecies are at risk.

JAVA SPARROW

THE JAVA SPARROW IS KNOWN the world over as a popular cage bird that breeds well in the right conditions. The adult bird is a little smaller than a house sparrow, *Passer domesticus*. The natural plumage is bluish gray with a pink belly and a black rump and tail. The head is black with large white cheek patches, and the bill is pinkish white. In captivity, other color variations have been bred, including pied, cinnamon and white.

The original home of the Java sparrow is on the islands of Java and Bali, but the species has been introduced into many parts of Southeast Asia. It is also called the Java finch, ricebird and Java temple bird.

Rice-eaters

Java sparrows have traditionally been a significant pest in Southeast Asia, where rice is the main crop, and their scientific name, *Padda oryzivora*, meaning "paddy field rice-eater," testifies to the fact. The birds forage in pairs or small flocks, stripping the ripening rice. They move on when the rainy season arrives to flood the rice paddies. Then the sparrows forage among bushes, feeding on fruit and snapping up any small insects they find. They also eat grass seeds and some maize.

Sociable sparrows

Highly social, the sparrows are often seen in large flocks in rice paddies, trees and thickets, though such flocks are much diminished today.

Pairs of Java sparrows live amicably together, sharing the same roost and greeting each other on meeting. If a captive pair is separated and then allowed to meet again, the birds perform an elaborate and charming ceremony. The first stage involves a low bow, with both partners trilling in unison. Then they perch side by side and each in turn twists sideways to rest its head over that of its mate. This is followed by more bowing, and the birds continue to sit very close to each other for some time.

Small flocks of Java sparrows gather to feed on rice, maize and other grasses. The rice thieves can be a pest to farmers.

The male Java sparrow's song is a fluting, sometimes whistling, tune that varies between individuals. It often ends with a long-drawn-out, metallic whistling note. Writers have given quite different descriptions of the song, which can be heard when the female is out of sight of the male and also during courtship. At the start of the courtship display, the male bows in a hunched posture; this is different from the "meeting" bow. He then hops up and down on the perch, waving his bill from side to side and twisting his tail toward the waiting female. As he nears her, he begins to sing. Sometimes the female bows and bounces as well, the pair of them making a pretty spectacle, before crouching with tail quivering. If the female rejects the male's advances, she may put her head over his, as in the meeting ceremony, or she may attack him, pecking at his bill.

Female finishes the nest

The nest is built from grasses and other plant materials that are long, flexible and tough. They are woven into a loose ball, and the chamber so formed is lined with feathers. In captivity the male has been seen to do all the nest-building. The female accompanies her mate on the trips he makes to collect material but helps him only at the end, when she collects feathers for the lining.

Both sexes incubate the eggs, taking turns of 20–30 minutes. The chicks hatch after about 3 weeks' incubation. They are completely naked at first and are brooded and fed by both parents.

Trouble at home

Unusually among birds, the Java sparrow is becoming very rare in many parts of its native range, while populations introduced elswhere in the world, such as China and Hawaii, thrive. Its scarcity in Indonesia is due to heavy persecution. Local people trap the birds for consumption, though many are exported to China for the table. Farmers wage a constant war against the rice thieves, although today they tend to chase the birds off their fields rather than shoot them.

This very pretty species is also a sought-after cage bird. During the mid-20th century tens of thousands were exported to the United States alone, and their reluctance to breed in captivity fueled the supply of wild-caught birds. The United States eventually banned their import in the early 1970s. Native populations may face further threats from the spread of the introduced Eurasian tree sparrow, *Passer montanus*, which is said to displace the Java sparrow where the two species now overlap in range. Various plans are in motion to protect the Java sparrow in its homeland. These include the establishment of preserves, a ban on wild capture and the promotion of captive breeding.

JAVA SPARROW

CLASS **Aves**

ORDER **Passeriformes**

FAMILY **Estrildidae**

GENUS AND SPECIES ***Padda oryzivora***

ALTERNATIVE NAMES
Java finch; Java temple bird; ricebird

LENGTH
Head to tail: 6–6½ in. (15–16 cm)

DISTINCTIVE FEATURES
Adult: very large, bright pink bill; black forehead, crown, nape, rump and tail; bright white cheeks and ear coverts; rest of plumage soft gray, with pinkish tinge to belly. Juvenile: grayish pink bill; lacks contrasting head pattern and pink on belly.

DIET
Cultivated rice and maize; also other small seeds, fruits and insects

BREEDING
Age at first breeding: 1 year; breeding season: June–July and December; number of eggs: 3 to 5; incubation period: about 21 days; fledging period: not known; breeding interval: probably less than 1 year

LIFE SPAN
Not known in wild

HABITAT
Crop fields, lowland grassland, open woodland with scrub, gardens and parks

DISTRIBUTION
Native range: Java, Bali and Bawean. Introduced range: India, mainland Southeast Asia, Borneo, Sulawesi, the Philippines, East Africa and Hawaii.

STATUS
Native range: generally scarce due to trapping and persecution. Introduced range: common in parts.

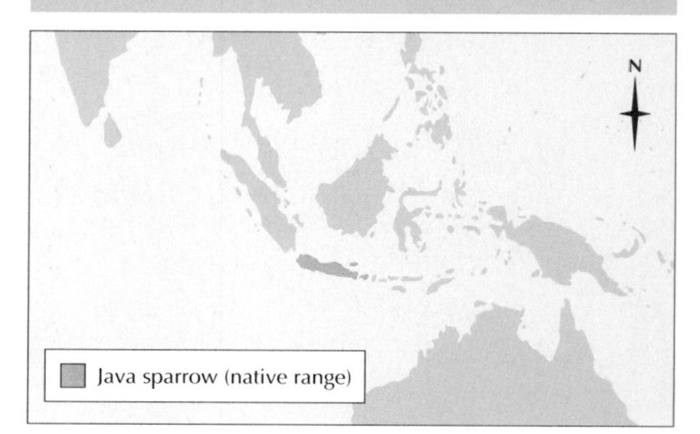

JAY

UNLIKE MOST OTHER members of the crow family, Corvidae, the jays are highly colorful. They make up a varied group of more than 40 species. About 30 species are found in the Americas, and most of these are South American. Eight jays occur in North America, including the blue jay (*Cyanocitta cristata*), Steller's jay (*C. stelleri*), gray or Canada jay (*Perisoreus canadensis*) and scrub jay (*Aphelocoma coerulescens*). However, the most cosmopolitan, or wide-ranging, species of jay is the Eurasian jay, *Garrulus glandarius,* which is discussed here in greatest detail. It occurs in most of Europe and in a broad band across Siberia to the Himalayas, China and Japan, as well as in North Africa.

Colorful crows

The Eurasian jay, about 13½ inches (34 cm) long, has a pinkish brown plumage, darker on the back and wings than on the breast. Most subspecies have a white crown streaked with black, and the feathers are raised in a crest in moments of excitement. There is a conspicuous black mustache running from the corner of the bill. The most noticeable feature is the patch of bright blue feathers barred with black on each wing. Eurasian jays tend to be shy, and the usual view of them is a black tail and pure white rump vanishing into the trees.

The blue jay of eastern and central North America is another common and beautiful species. It is 11 inches (28 cm) long, blue on the upperparts and whitish gray on the underparts, with black bars on the wings and tail and a black breast band. A pointed crest gives the blue jay a distinctive silhouette. Steller's jay is similar but is blue all over, with a blackish crest and throat. It replaces the blue jay in western North America, including in the Rockies (the blue jay occurs only in the eastern Rockies).

The Eurasian jay is the most colorful species of crow in Europe. Its screeching call is a familiar sound in deciduous woodland.

Eurasian jays collect and bury hundreds of acorns in autumn, to last them through the winter months. The jays' remarkable visual memory enables them to relocate most of the hidden stores of food.

The gray jay has a soft gray and white plumage, and is found mainly in Canada and the Rockies. Similar in shape but with a brownish red plumage, is the Siberian jay, *Perisoreus infaustus*. It lives in Scandinavia and across northernmost Asia. The green jay, *Cyanocorax yncas*, is a tropical species found in Central and South America, reaching as far north as the southern tip of Texas. It has bright green upperparts blending into blue on the head, a pale green belly and a bright yellow patch under the tail. Although garish, this plumage blends in well with the dappled sun and shade of tropical forests and bush country.

Shy but garrulous

The Eurasian jay lives in woods and forests but comes out onto the open scrubland nearby to feed. Highly secretive, its harsh cries can be heard among trees for long periods without the bird showing itself. Its flight is heavy and appears to be labored. The raucous cries and catlike mewing are the best known of its calls, but it uses a very soft and melodious song in spring. Eurasian jays are also good mimics, and a tame jay can mimic a wide range of mechanical sounds, bird songs and human speech. Even wild jays use vocal mimicry. For instance, one individual always hooted like a tawny owl, *Strix aluco*, when it flew past the tree in which an owl was known to roost.

EURASIAN JAY

CLASS **Aves**

ORDER **Passeriformes**

FAMILY **Corvidae**

GENUS AND SPECIES ***Garrulus glandarius***

WEIGHT
5–6⅔ oz. (140–190 g)

LENGTH
Head to tail: 13⅓–13¾ in. (34–35 cm); wingspan: 20½–23 in. (52–58 cm)

DISTINCTIVE FEATURES
Stocky, compact body; large head; stout bill; broad wings; strong legs; white crown with fine black streaks; black mustachelike stripe from corner of bill to cheeks; pink, pinkish gray or pinkish brown back and underparts; bright blue patch on each wing; pure white rump (visible in flight only); black tail

DIET
Mainly invertebrates, fruits, seeds and nuts, particularly acorns; occasionally small vertebrates, bird eggs, carrion and scraps

BREEDING
Age at first breeding: usually 2 years; breeding season: eggs laid April–May; number of eggs: usually 5 to 7; incubation period: 16–17 days; fledging period: 21–22 days; breeding interval: 1 year

LIFE SPAN
Up to 18 years

HABITAT
Mainly deciduous woodland and forest; also parks, large gardens and orchards

DISTRIBUTION
Much of Europe, south to North Africa and east through Siberia to China and Japan

STATUS
Common

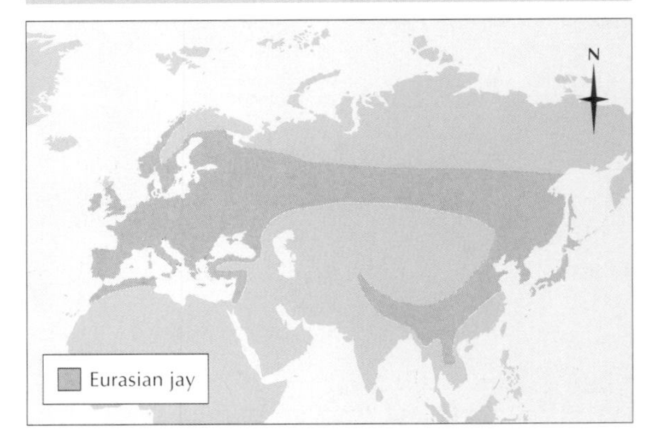

Other species of jay are also woodland birds. The gray and Siberian jays inhabit the northern coniferous forests of the boreal climatic zone. Steller's jay favors pine-oak forests and coniferous forests. The blue jay lives in a variety of wooded habitats, including suburban parks and gardens. Gray and Steller's jays are also familar campsite visitors, boldy snatching scraps of food from around campfires and tents. Meanwhile, the scrub jay, as its name suggests, occurs in areas of scrub oak, chaparral and open pinyon-juniper woods.

Caches of buried food

The Eurasian jay's diet is very varied and includes seeds, fruits and small invertebrates such as insects, spiders, earthworms, snails and slugs. It also takes the eggs and nestlings of small songbirds, wood pigeons and pheasants. In autumn the main food is acorns, and large quantities are eaten. A young jay, hand-reared and isolated from its own kind, will strip an acorn of its husk in a way that suggests it knows instinctively how to deal with it.

Jays also bury acorns, pecking a hole in the ground, placing the acorn in it and then covering it up. Tests have shown that, surprisingly, jays can find these acorns again. They seem to remember where each acorn is buried and will return unerringly to it even when the spot is covered with up to 15¾ inches (40 cm) leaf litter or snow.

Spring gatherings

In spring Eurasian jays indulge in what are called ceremonial assemblies, the significance of which is not yet clear. Groups of three to 20 birds chase each other from tree to tree or among the branches of the trees uttering a chorus of soft warbling notes, which are delightful to the ear and in strong contrast to their usual harsh notes. It has been suggested that these assemblies may have something to do with the birds pairing off but this has not been confirmed.

The breeding season of the Eurasian jay is April–May, when a nest is built 5–20 feet (1.5–6 m) from the ground in a tree. It is made of sticks with a little earth and is lined with fine roots. In this five to seven eggs are laid, green to buff colored with fine dark mottlings, and scrawled with fine hairlines. Probably both parents incubate for 16–17 days, and the young are fed for about 3 weeks after hatching. It has yet to be proven that the two parents share the incubation. Male and female look alike and can only be distinguished by their behavior when courting.

Communal breeding

The isolated Florida population of the scrub jay is of great interest to ornithologists because this subspecies has evolved a communal breeding system. Communal breeding refers to a pair of birds (male and female) being assisted in their breeding by one or more helpers. These helpers are usually offspring from previous breeding seasons that do not yet have a territory of their own. They feed the young and help to chase away predators such as snakes. Because the helpers are related to the brood of young, they share much of the chicks' genetic make-up. It is better to help raise them than to not breed at all.

***A pair of blue jays. The most common call of this noisy species is a piercing* jay jay jay.**

JELLYFISH

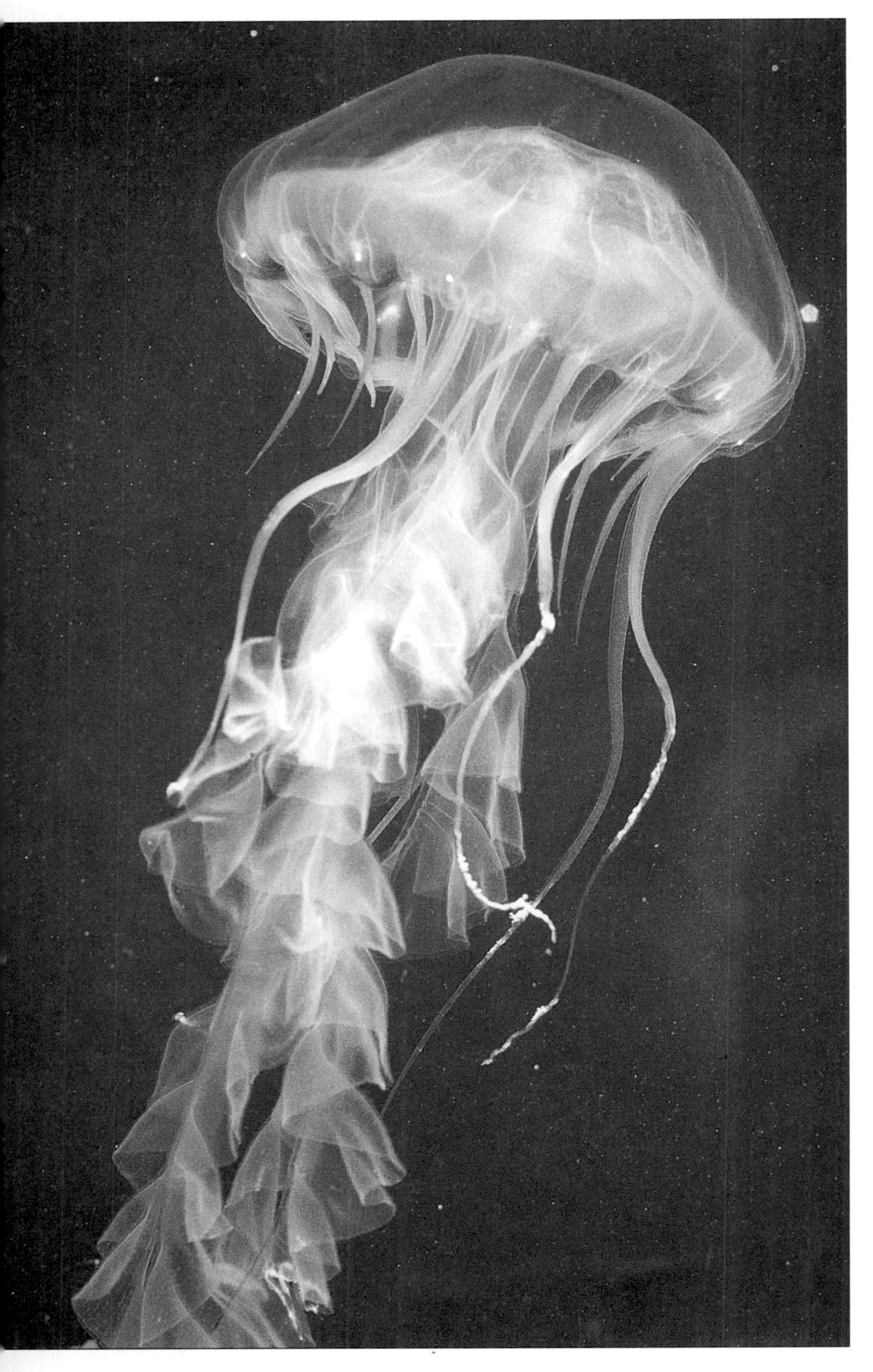

***Some of the largest jellyfish belong to the genus* Chrysaora (C. hyoscella, *above*).**

THE JELLYISH ARE FREE-SWIMMING relatives of sea anemones, corals and hydras, all belonging to the phylum Cnidaria. There are two distinct phases in the life cycles of many cnidarians. In one of these a free-living jellyfish, or medusa, reproduces sexually. In the other the organism develops from an embryo and is a sessile (anchored) polyp, or colony of polyps, which in turn buds off jellyfish. One phase or the other may be dominant and the other less important or even absent. The large jellyfish make up one class, the Scyphozoa (or Scyphomedusae), in which the polyp stage is very small. Attention is concentrated here on this group. Although they can occur in high numbers, there are relatively few Scyphozoan species, about 160 in all.

Trailing tentacles

The typical jellyfish is umbrella-shaped, globular or conical with four or eight tentacles around the margin, or many tentacles forming a ring around the margin. Under the umbrella and extending from it is the mouth, leading into the digestive cavity. The mouth is drawn out at the corners into four long lips. The basic form of the body can best be understood by comparison with that of hydras. The body of a hydra consists essentially of two layers of cells forming a sac and separated by a very thin layer of noncellular material, mesogloea. In the jellyfish the mesogloea is very thick. Although the body of a jellyfish is more elaborate than that of a hydra, it still has the same two-layered structure and ring of tentacles around the mouth.

Some common jellyfish

A jellyfish that is found in seas throughout the world and is common off the coasts of Europe is *Aurelia aurita*. It grows to nearly 1½ feet (45 cm) across, with many very short tentacles. The blue, yellowish, reddish or brown jellyfish *Cyanea*, also known as sea blubber or lion's mane, can reach 6 feet (1.8 m) across in Arctic waters but is usually less than half that. Jellyfish of the genus *Chrysaora* have 24 tentacles, and these may be 60 feet (18 m) long in one species. Around the center of the white or yellowish disc there is often a brownish ring from which streaks of the same color radiate. Another common jellyfish is *Rhizostoma*, or "the root-mouthed," named for the shape of its lips. It is a whitish dome, about 1 foot (30 cm) across, with a deep purple rim. It has no tentacles but is easily recognized by the cauliflower-like oral lips. In the United States it is called "cabbage blebs." Some jellyfish are luminescent, and one of the most intense, which strays to European waters, is *Pelagia noctiluca*.

Different ways of feeding

Jellyfish swim by rhythmic pulsations of the umbrella, or bell. The movement resembles that of an umbrella being opened and shut slowly.

It is coordinated by a simple nervous system and by sense organs around the edge that are sensitive to light, gravity and chemicals in the water.

Hunting the plankton

Jellyfish are carnivorous, and many of them capture fish, shrimps and other animals on their trailing tentacles. They paralyze the prey with their stinging cells and transfer it to the mouth. *Aurelia* feeds on small planktonic animals. These are trapped in sticky mucus all over the surface of the body and are driven by cilia (tiny hairlike extensions) to the rim. There the accumulating blobs are licked off by the four long oral lips. Further cilia propel the food in currents of water into the digestive cavity, from which a system of branching, cilia-lined canals radiate out to the rim, carrying food to all parts of the body.

Rhizostoma feeds in the manner of a sponge, drawing in small planktonic animals by means of ciliary currents through thousands of separate mouths on the greatly elaborated oral lips. It is these mouths and the many little sensory palps on the oral lips that give the jellyfish its characteristic cauliflower appearance. *Cassiopeia* is a tropical plankton-feeder that lies mouth upward on the sea bottom in shallow water, pulsating its bell gently and capturing the wafting plankton with its lips. It has symbiotic algae in its oral lips, which benefit from the sunlight that falls on them from above. The likely significance of jellyfish in controlling numbers and types of zooplankton is only now beginning to be appreciated by scientists.

MOON JELLYFISH

PHYLUM **Cnidaria**

CLASS **Scyphozoa**

ORDER **Semaeostomea**

FAMILY **Ulmaridae**

GENUS AND SPECIES ***Aurelia aurita***

ALTERNATIVE NAMES
Moon jelly; common jellyfish

LENGTH
Bell (main body): usually less than 10 in. (25 cm) across; rarely up to 18 in. (45 cm)

DISTINCTIVE FEATURES
Umbrella-shaped bell; 4 frilly feeding lips; many short tentacles; 4 nearly oval purple reproductive organs clearly visible inside bell, otherwise almost completely transparent

DIET
Plankton up to ⅕ in. (5 mm) in diameter, including diatoms, algae, invertebrate eggs, young polychaete worms, mollusk larvae and copepods; some larger plankton and very small fish

BREEDING
Sexes separate. Breeding season: summer; tiny planula larvae released and then settle, developing into colonies of polyps; each polyp buds off and develops into jellyfish.

LIFE SPAN
Up to 1 year

HABITAT
Coastal waters; sometimes also estuaries

DISTRIBUTION
Virtually worldwide in Pacific, Atlantic and Indian Oceans

STATUS
Abundant; many populations undergo large seasonal variations

Piles of saucers

The common *Aurelia* is readily recognized by the four nearly oval purple or lilac reproductive organs—ovaries in the females, testes in the

***Jellyfish vary in form but typically the bell (main body) is shaped like an umbrella or cone, below which is a variable number of tentacles. Pictured is* Cephea cephea.**

gives rise in the following winter to a number of young jellyfish called ephyra larvae. These are not round like the adult, but with the edge of the bell drawn out into eight arms, notched at the tips. To do this, the polyp becomes pinched off into segments so that it resembles a pile of lobed saucers. Then the tissue connecting these saucers contracts and snaps, and each one swims off as a little ephyra. The growing ephyras gradually transform into adults by filling in the spaces between the arms. An alternation of forms such as this is typical of these jellyfish, although in *Pelagia* the egg develops directly into an ephyra.

Floating menace

Jellyfish are practically all water. A jellyfish stranded on the shore will soon vanish under the hot rays of the sun, leaving little more than a patch of water in the sand. Their bodies are 95–99 percent jelly, and the whole body contains less than 5 percent organic matter. Yet jellyfish can be extremely venomous, as anyone knows who has hauled on a rope covered in long, trailing tentacles.

The stings of jellyfish come from the many stinging cells, or nematocysts, which shoot out a venom-laden thread when touched. The severity of the sting depends very much on the number of nematocysts discharged and on the type of jellyfish. The most venomous jellyfish are those living in the coral seas, and the least troublesome are those living in temperate seas, but even these, if enough tentacles are allowed to touch the body, can sometimes lead to a loss of consciousness and, in rare cases, to drowning.

A jellyfish's mouth is drawn at the corners into oral lips, which are long and frilly in many species. The lips trail beneath the main body, surrounded by the much thinner stinging tentacles.

males. The organs lie in pouches in the digestive cavity but show through the transparent bell. The male sheds his sperm into the sea, and these are wafted to the female and taken in along with her food. The eggs are fertilized and develop for a while in pouches on the oral lips. They are eventually set free as tiny planula larvae, which soon attach themselves to seaweed, stone or any other suitable surface and develop into small polyps, known as scyphistomas or hydratubas, each with 16 tentacles.

From the base of each polyp, stolons, like runners of strawberry plants, grow out and new polyps develop on them. Each polyp eventually

Sea wasps

The most venomous jellyfish belong to the Cubomedusae, named for their squarish shape. They range in size from that of a grape to that of a pear, and have four tentacles or four groups of tentacles. Some favor quiet, shallow waters in the warmer seas and are feared by divers and bathers around the tropical northern Australian coasts, the Philippines and Japan. Most lethal of all is the box jellyfish, *Chironex fleckeri*. A brush with its tentacles raises red welts on the skin and causes instant, excruciating pain. A major sting can cause cardiac dysfunction or even arrest within as little as 15–40 minutes.

JERBOA

JERBOAS ARE SMALL to medium-sized rodents, many of them inhabitants of desert regions. They look rather like kangaroos, having long hind legs, very short forelegs and long tails. Like kangaroos, they travel by hopping. Other desert-living rodents have developed hopping as a means of traveling across sparsely covered ground, but in none are there such great differences between forelegs and hind legs as there are in the jerboas, whose hind legs are four times longer than their forelegs.

The tail can be longer than the head and body and often bears a white tuft at the tip. It serves as a balancer when hopping and as a prop when the jerboa is sitting upright. The fur is fine and usually sandy in color, typically matching the local ground color. Some species have long, rabbitlike ears, and others have short, mouselike ears. Jerboas are found in northern Africa and Asia east to northern China and Manchuria.

Avoiding the heat

Desert-dwelling jerboas have long ears to radiate excess body heat. They spend the day in burrows and emerge in the cool of night to forage. Their burrow entrances are usually found near vegetation, especially along field borders, but during the rainy season they tunnel into mounds or hillsides where there is less risk of flooding. At this time of year the entrances to the burrows are left open. During the summer occupied holes are plugged, to keep out hot air and perhaps predators, too. Burrows often have an emergency exit that ends just below the surface or opens at the surface but is loosely plugged. When disturbed, a jerboa can burst through this exit and flee. Asian jerboas dig winter burrows, 10 feet (3 m) or more long, which may be plugged and are consequently difficult to trace.

A jerboa burrows very rapidly, scraping the soil with its sharp-clawed forefeet and kicking it back with the hind feet. Even the nose is used for bulldozing soil or tamping the burrow walls. Some species have a fold of skin on the nose that can be drawn over the nostrils to keep out sand.

When chased, jerboas race away at up to 15 miles per hour (24 km/h), covering 10 feet (3 m) at each bound. Otherwise they "trot," leaping 4–5 inches (10–12 cm) at each bound. When foraging, they shuffle on all fours.

Feed on succulent plants

Jerboas feed mainly on plants. In the desert during the rainy season there is plenty of fresh greenery. As the plants die back, the jerboas dig out roots in which water is stored, and in the dry season they survive on dry seeds. In some places jerboas attack crops of watermelons and rubber trees, and certain species feed on beetles and other insects. The jaws of jerboas are weaker than those of gerbils, so they cannot eat hard seeds.

A jerboa's bounding gait enables it to cover a large area in search of food with the minimum expenditure of energy. This is vital in the dry season when food is scarce. Unlike gerbils, jerboas do not store food.

Huge hind legs enable jerboas to dig tunnels and to leap away from danger with great speed and efficiency. However, they usually travel at a sedate pace.

Slow-growing babies

Most jerboas are solitary animals, each adult having its own burrow and foraging alone, although sometimes loose colonies are formed. Some species appear to have communal burrows, which offer extra warmth during cold weather.

Jerboas tend to be solitary animals, each adult having its own burrow and foraging alone. They emerge after sundown, during the cool of night.

Breeding probably takes place twice a year in most species and some jerboas may have more than two litters, but the young of the last litter may die before reaching maturity if they are born too near the start of the dry season. Litters are of two to six, but usually three, babies. Naked at birth, the babies crawl with their forelegs, since the hind legs do not develop until 8 weeks; it is another 3 weeks before the young can jump. Weaned at 8 weeks, they are sexually mature at about 14 weeks.

Escape at top speed

Unlike other rodents, jerboas do not dash for their burrows or other cover when pursued but bound off at high speed, with frequent changes of direction and the occasional high leap. They are probably on the menu of all local predators, such as foxes, owls and snakes. Bedouin Arabs have traditionally caught them by flooding or digging out their burrows, or by setting traps.

A dry diet

By living in a cool, humid burrow and emerging at night jerboas can escape the worst of the desert heat, but conditions are still severe by comparison with those in temperate climates and, like other desert animals, jerboas face the problem of water shortage. They economize on water by living in burrows, and in the hottest weather (and winter, too) they become dormant. So efficient are they at conserving body fluids that they can survive the summer on dry seeds containing almost no moisture. In laboratory conditions jerboas have lived for 3 years on a diet of dry seeds. By comparison, rats could survive for only a few days. The jerboas' survival on this diet is due to their ability to retain water in the body and pass a concentrated, acidic urine, although in such extreme circumstances jerboas are much less active, so less body waste is formed and fluid retention is easier.

JERBOAS

CLASS **Mammalia**

ORDER **Rodentia**

FAMILY **Dipodidae**

GENUS **14, including *Allactaga, Dipus, Eozapus, Euchoreutes, Jaculus, Paradipus, Salpingotus* and *Stylodipus***

SPECIES **34 species**

ALTERNATIVE NAMES
Jumping rat; kangaroo rat (unrelated to American kangaroo rats, genus *Dipodomys*)

WEIGHT
1–4 oz. (28–120 g)

LENGTH
Head and body: 1½–6¼ in. (3.5–16 cm); tail: 2¾–5 in. (7–13 cm)

DISTINCTIVE FEATURES
Hind legs much longer than forelegs; fine, dense fur; sandy to dark brown upperparts; white underparts

DIET
Seeds, leaves, shoots and insects

BREEDING
Age at first breeding: usually 2 years; breeding season: usually summer; gestation period: 90–120 days; number of young: usually 2 or 3; breeding interval: 6 months

LIFE SPAN
Up to 2–5 years

HABITAT
Desert, sand dunes, scrub and grassland

DISTRIBUTION
North Africa; Middle East; south-central Asia east across southern Russia, Mongolia and northern China

STATUS
Most species rare; 12 species threatened, vulnerable or endangered

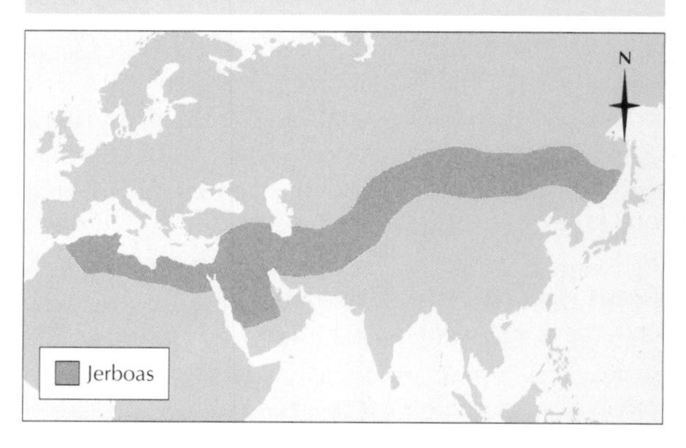

JEWEL FISH

The jewel fish is one of the most beautiful of African cichlids and a favorite with aquarists as well as with those studying fish behavior. The adults are gorgeously colored and seem to be spangled with jewels. They are up to 6 inches (15 cm) long, and the coloration is similar in both male and female, although at times the female is the more brilliant. The male has, however, more "jewels," especially on the gill covers, and a more pronounced crescent on the tail fin. Jewel fish vary greatly in color and pattern according to the time of year and the mood of the fish. The base color is dark olive to gray-brown with a greenish sheen, underlying various spots and blotches. During the breeding season the underside is a brilliant lacquer-red.

The jewel fish is found in rivers over most of tropical Africa, from the Gambia, Niger and Congo east to the upper Nile, southeast to Lake Tanganyika, and across to eastern Sudan.

Colorful start to life

As the time for egg-laying draws near, the red body coloration intensifies and extends. Male and female spend much time lying side by side on the sandy bottom during a period of two or three days, and then begin to clean a hard surface for the eggs. The female lays her 1-millimeter eggs in rows that look like short strings of tiny pearls. The male follows her to fertilize each row, until a rounded patch of 500 to 700 eggs covers the area. The female fans the eggs with her pectoral fins to aerate them, with the male taking over when she forages.

The baby fish begin to feed at about 7 days old, first on protozoans and then on rotifers (minute invertebrates) and small crustaceans. After 30 days they resemble the adults. Although still tiny, they start to fight, displaying a pugnacious quality common to many cichlids. They grow ¾ inch (1.8 cm) each month.

A color key for mothers

The bright colors of the jewel fish play an important role in life. They protect it as a baby and help it find a mate. The babies are herded by their parents as soon as they swim, for if abandoned they would soon be eaten. The parents clearly have some way of calling their broods to them when danger threatens, and the babies must have some means of recognizing their parents' warning signals. Experiments in captivity have shown that color plays a role in keeping the baby fish with their parents.

In early tests, three glass aquaria were placed side by side with their long sides touching. Baby jewel fish were put in the middle tank, and in each of the other tanks was put a disc on a long rod, the rods being fixed above on a converted windshield wiper. When this was set in motion, the discs moved back and forth in sight of the baby fish. One disc was painted scarlet, the other black. As soon as the discs started moving, the young jewel fish moved over to the side of their aquarium nearest the scarlet disc Biologists then used discs of various sizes, 1–3 inches (2.5–7.5 cm) in diameter, and different colors. They experimented with the speed of oscillation, and also experimented with broods of different ages.

The tests revealed that baby jewel fish have an innate preference for scarlet, the mother's main body color, over all other colors. Moreover, this preference grows even stronger as the babies grow. After several weeks, however, their liking for red declines, signaling the time to disperse. Meanwhile, the parents' bodies lose their red coloring, and eventually the family breaks up.

Size is important

The experiments also revealed that a baby jewel fish favors discs of 2–3 inches (5–7.5 cm) diameter, about the size of their parents, over smaller

During the breeding season the jewel fish is covered with sparkling blue spots, and its belly turns a brilliant red. It is one of the most colorful of all the African cichlids.

A female jewel fish guarding her 3-day-old young. Baby jewel fish start to feed at 7 days, and at 30 days are already aggressive toward one another.

ones. Movement also makes a difference. A motionless scarlet disc barely attracts the babies. In the same way, a sluggish parent will not rally its brood effectively. An active parent, sensing danger, moves more rapidly and so imparts a sense of urgency to its brood. Moreover, at such times the mother raises and lowers her unpaired fins several times in succession to lure her babies.

Tests on courting fish show similar results. For example, a jewel fish is more attracted to a partner that not only shows red but also moves quickly than to one that shows even more red but moves sluggishly. The female clearly favors a mate that moves quickly: he will be much more likely to protect her eggs.

No two alike

The detailed color pattern of the jewel fish has its value as much as the general color. Once two fish have paired, they recognize each other even when among a crowd of their own kind, which to our eyes look identical. A male jewel fish knows his own mate even when there are several females in his territory. He drives the others away but does not molest her; and tests have shown that he recognizes her by subtle differences in the pattern of her colors. When the male of a pair is taken out of an aquarium and one or more strange males introduced, the female attacks them and tries to drive them out. However, she welcomes her mate when he is replaced in the aquarium.

There have been instances in which a female, after mating with a male, sees him again the following year in a nearby aquarium. She attempts to get to him even though another male has been put into her aquarium.

JEWEL FISH

CLASS **Osteichthyes**

ORDER **Perciformes**

FAMILY **Cichlidae**

GENUS AND SPECIES ***Hemichromis bimaculatus***

ALTERNATIVE NAMES
African jewel fish; jewel cichlid

LENGTH
Up to 6 in. (15 cm)

DISTINCTIVE FEATURES
Elongated, laterally compressed body; highly variable coloration.
Nonbreeding adult: dark olive to gray brown overall, with greenish sheen; greenish yellow flanks; yellowish belly.
Breeding adult: brilliant red underside; olive green forehead and back, with reddish sheen; dark blue blotch on both gill cover and flanks; many sky blue spots or "jewels" on much of body.

DIET
Adult: most organic matter, from submerged plants to crustaceans and mollusks.
Young: small invertebrates.

BREEDING
Age at first breeding: when 3 in. (7.5 cm) in length; number of eggs: 500 to 700; breeding interval: variable

LIFE SPAN
Not known

HABITAT
Near bottom of freshwater and brackish rivers and canals; often associated with forested regions

DISTRIBUTION
Tropical and subtropical Africa

STATUS
Common

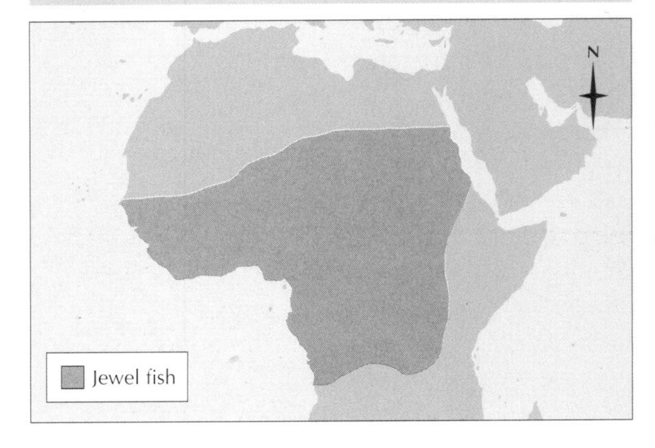

JOHN DORY

THE PECULIAR-LOOKING FISH called the John Dory is included in an order known formerly as the Zeomorphi but now called Zeiformes. Literally translated from the Greek, these terms mean "godlike" or "god-shaped" respectively. The scientific name, *Zeus faber*, continues the divine theme. It means "blacksmith of Zeus" (the overlord of the Greek gods).

Flattened body

With its high and narrow body, flattened from side to side and oval in outline, the John Dory is shaped rather like a serving dish. The large head has a mournful expression due to the drooping mouth, and the jawbones are formed in such a way that the mouth can be shot forward a short distance to assist in seizing prey. The dorsal fin is in two parts. The front portion is tall, with strong spines. In older individuals the spines are long and carried backward to align with the tail. The rear part of the dorsal fin and the anal fin that opposes it are soft and flexible. Along the bases of the dorsal and anal fins are spines, but the scales on the body are small and spineless, and the skin is smooth. There are also eight or nine spiny plates along the belly. The John Dory is gray to golden yellow overall with long blotches of reddish purple, and on each flank behind the gill covers is a black spot with a yellow margin.

The John Dory is found fairly close to the shore. It ranges from the coasts of western and southern Europe, including the Mediterranean, around Africa and Asia to Australia and New Zealand. It has been recorded as far north as Norway in the North Atlantic.

Catlike stalking

With its platelike body, the John Dory is not adapted for chasing after prey. Instead, it stalks its quarry, keeping its body rigid and swimming by waving its second dorsal and its anal fin. The tail fin serves as a rudder. The flattened body helps in enabling the fish to slip easily through the water for short distances. Seen from head-on, the high and narrow body appears as a slender, vertical stripe. Any small fish being stalked is largely unaware of the impending danger and makes little attempt to escape. Keeping its eyes

The John Dory stalks its prey rather than giving chase. It feeds mainly on schooling fish.

The John Dory's body is strongly compressed from side to side, and this makes the fish very inconspicuous when seen from the front or back.

JOHN DORY

CLASS **Osteichthyes**

ORDER **Zeiformes**

FAMILY **Zeidae**

GENUS AND SPECIES ***Zeus faber***

ALTERNATIVE NAMES
Dory; doree; St. Peter's fish; girty

WEIGHT
Up to 6½ lb. (3 kg)

LENGTH
Up to 3 ft. (0.9 m)

DISTINCTIVE FEATURES
Very high, deep body with large head, so laterally compressed as to be inconspicuous when viewed head-on; protrusible jaws; small spiny plates at base of dorsal and anal fins; long ventral fins; yellowish to gray

DIET
Mainly other fish; occasionally cephalopods and crustaceans

BREEDING
Breeding season: June–August (northeastern Atlantic); breeding interval: 1 year

LIFE SPAN
Up to 12 years

HABITAT
Near bottom in coastal and brackish waters

DISTRIBUTION
Eastern Atlantic, including Mediterranean and Black Sea; Indian Ocean; western Pacific

STATUS
Common

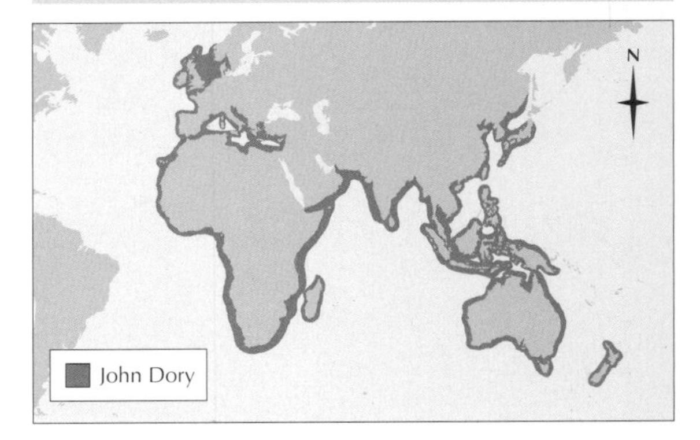

on its prey, the John Dory gradually closes in, finally seizing the victim by shooting out its protrusible, toothless jaws.

The John Dory is a secretive fish, and what little is known about its lifestyle comes mainly from aquarium studies. It feeds almost wholly on small fish, especially young herring, pilchards and sand eels, although in captivity it has been seen to take shrimps.

There is no outward difference between the sexes. Only by dissection and examining the sex organs is it possible to distinguish female from male. In the northeast Atlantic the eggs are laid between June and August. Tiny globules of oil in the eggs help them float among the plankton.

Dual personality

People are divided as to the origin of the John Dory's name. Was John Dory perhaps the name of a swashbuckling buccaneer? The species is also surrounded by legend. One bible-centered tale asserts that the black blotches on its flanks mark where the finger and thumb of St. Peter pressed when he took a coin from the fish's mouth to pay his tax. This accounts for the name *Peterfisch* in Germany.

JUNGLE FOWL

The four species of jungle fowl are highly distinctive members of the pheasant family. In each species the cock (male) has fleshy appendages on his head: a comb rises from his crown, and lappets hang from the chin, while most of the face is naked. In the hen (female) the appendages are vestigial (imperfectly developed) and the plumage is duller. The best known species, the red jungle fowl, *Gallus gallus,* is the ancestor of the domestic chicken. Its plumage is mainly red and greenish with an iridescent sheen.

The red jungle fowl is native to warmer parts of Asia, from the Indian subcontinent to southern China and the Indonesian islands. The populations on various Pacific islands, such as the Philippines and Borneo, were probably introduced by humans. The red jungle fowl has also been established in Natal, South Africa, and there are free-living populations in several parts of the world. The gray jungle fowl, *G. sonnerati,* lives in the southern and western parts of the Indian peninsula, and the green jungle fowl, *G. varius,* lives in Java. In the fourth species, *G. lafayetii,* which is confined to Sri Lanka, the cock's comb is yellow with a red border. Its plumage is red and brown with greenish black iridescent wings and tail.

A male green jungle fowl, Java, Indonesia. Jungle fowl scratch for food among the soil and leaf litter, rather like domestic chickens.

Keeping out of the way

Jungle fowl are forest-dwellers. Their range is shrinking as forests are felled and the terrain is given over to agriculture. Their extreme wariness undoubtedly contributes to their survival. Jungle fowl emerge to feed in clearings, roadsides and fields, especially in early morning or after rain, but run for cover at the first alarm. Even when undisturbed, they remain alert.

In the bamboo thickets of Thailand, jungle fowl live in small flocks of one cock and two to five hens. In the morning the flock leaves its roost 15–20 feet (4.5–6 m) up in the bamboo stems and is led down by the cock to drink at a stream. While the hens drink, the cock keeps watch from a nearby perch. When they are finished, the cock drinks hurriedly and then quickly leads his hens back to cover.

Scratching for food

Jungle fowl eat virtually anything they can find in the leaf litter and soil of the forest floor, including green plants, seeds, berries, earthworms and insects. They forage like domestic chickens, scratching vigorously with their stout toes and stepping back to search for anything brought to light. In Sri Lanka the nellu plant forms a great part of the undergrowth in the forests. It flowers at intervals of several years, and large numbers of jungle fowl and other birds gather to feast on the nutritious seeds.

Cocky challengers

The breeding season varies across the species' range, in India coinciding with the March–May dry season. It brings out the aggressive side in cocks. Unmated males, which at other times loiter unobtrusively on the territories of mated cocks, now grow more daring. They challenge the residents by crowing and clapping their wings. The challenge cannot be ignored. Fierce fights break out in which the combatants may inflict deep wounds with the sharp, 1-inch (2.5-cm) spurs on their heels.

The cock courts the hen by waltzing around her with one wing lowered, rubbing the primary feathers of that wing with the nearest foot to produce a rasping sound. The hen makes a nest by scraping a hollow in the soil. Alternatively she uses the crown of a tree stump or takes over the abandoned nest of a large bird. She incubates her five or six speckled eggs for 18–20 days. The chicks can walk and feed themselves soon after hatching. They can fly weakly when a week old, when they begin to roost in the trees.

The domestic chicken shares its scientific name, Gallus gallus, *with the red jungle fowl (above), which is its wild ancestor. Domestication took place at least 3,000 years ago in Asia.*

Safe perches

Jungle fowl fall prey to all manner of meat-eaters. Their only defense is to be extremely wary and to take flight at the first alarm. The exception is the incubating hen, who stays on the nest until the last moment. Jungle fowl choose their night-time roosts with great care. They select perches as far out as possible on a tree limb, where the foliage shelters them from owls and the branches are too slender to support even such nimble creatures as the palm civets (*Paradoxurus spp.*).

Rapid turnover

The jungle fowl has been little observed in the wild, but scientists at San Diego Zoo were able to study the breeding strategies of the semiwild flocks roaming the site. Some of their findings were startling. For example, only six out of every 100 chicks survived the first year of life. Seventy-five percent died before reaching independence, and jungle fowl that survived to maturity did not live beyond three years. Foxes and cats were probably responsible for many of the deaths, so it seems that the jungle fowl's wariness and care in selecting a roost are insufficient protection.

A rapid turnover of population is found in many species and has certain advantages of its own. For example, it enables a population to rebuild rapidly when breeding conditions are ideal. That said, genetically pure jungle fowl are now said by some scientists to be scarce or facing extinction in the wild as a result of interbreeding with domestic or feral chickens. Habitat loss is an additional pressure.

RED JUNGLE FOWL

CLASS **Aves**

ORDER **Galliformes**

FAMILY **Phasianidae**

GENUS AND SPECIES ***Gallus gallus***

ALTERNATIVE NAME
Wild jungle fowl

WEIGHT
Male: 1½–3 lb. (670–1,450 g). Female: 1–2⅓ lb. (485–1,050 g).

LENGTH
Head to tail: male 26½–30 in. (67–75 cm); female 16½–18 in. (42–46 cm)

DISTINCTIVE FEATURES
Resemble domestic chicken. Male: red face; fleshy red comb (on crown) and lappets (below chin); long, maroon, rufous or golden neck feathers; glossy green wing coverts and tail. Female: pinkish face; brown upperparts, breast and upper belly; short, black and golden buff neck feathers.

DIET
Seeds, berries, shoots and invertebrates

BREEDING
Age at first breeding: 1 year; breeding season: variable; number of eggs; 5 or 6; incubation period: 18–20 days; fledging period: about 7 days

LIFE SPAN
Up to 3 years

HABITAT
Forest margins, open woodland and scrub

DISTRIBUTION
Eastern India east to southernmost China, south to Java, Indonesia

STATUS
Uncommon; some scientists claim pure-bred birds facing extinction in the wild

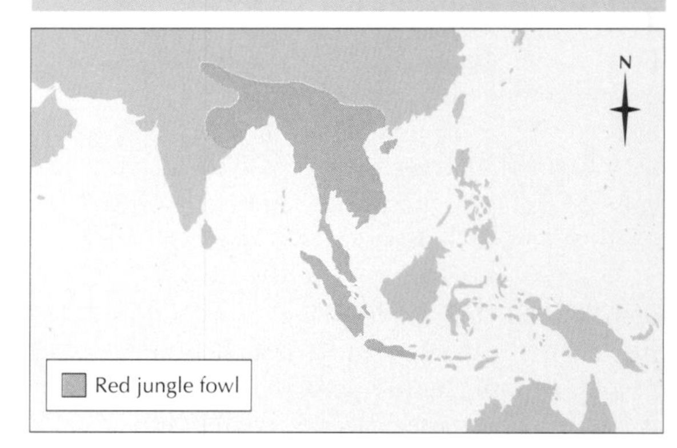

KALLIMA BUTTERFLIES

KALLIMA, FROM A GREEK word meaning beautiful, is the generic name of certain butterflies belonging to the family Nymphalidae. This same family includes the fritillaries, purple emperor and white admiral, and the vanessid butterflies such as the peacock, red admiral, tortoiseshell and Camberwell beauty, all brilliantly marked and powerful in flight. Kallima butterflies are also called leaf or, more commonly, dead-leaf butterflies. They too are colorful, strong fliers, but upon closing their wings they are transformed. The several species of Kallima range from New Guinea through Southeast Asia and southern Asia to India and Sri Lanka. Some are found in tropical Africa.

The Indian and Far Eastern species, *Kallima inachus* and *K. paralekta*, are 3½ inches (9 cm) across the spread wings. The upper side of the wings is patterned with dark brown, blue and bright orange, but in the Sri Lanka species, *K. philarchus*, the orange is replaced with white. The other species are variously colored but all have a similar kind of wing patterning.

Bogus foliage

In the kallima butterflies the shape of the wings when closed over the back, together with the colors and pattern of their undersides, gives the appearance of a dead leaf. Many members of the family have "tails" on the rear margins of the wings. These are short and blunt-tipped. When a dead-leaf butterfly alights on a twig, the wings fold over the back and form a "stalk" shape. The tip of the leaf is represented by the pointed, curved tips of the forewings as they lie together. Between this tip and the bogus leaf stalk runs a dark line, across both forewings and hind wings, which looks exactly like the midrib of a leaf.

Trembling like a leaf

Less distinct dark lines run obliquely upward from the central "midrib" line to the margins of the wings, and these look exactly like the veins of a leaf. To complete the illusion, and this is especially true of *K. inachus*, the species most often seen in museums or books, there are patches on the wings just like the holes and tears, the fungal growths and other blemishes found on dead leaves. The abdomen, thorax, head and antennae are inconspicuous when the butterfly is at rest, and the whole effect is such that once the butterfly has settled it is almost impossible to detect among the foliage. No two butterflies of the same species are patterned exactly alike on their underwings, just as no two dead leaves are exactly alike. As soon as the butterfly settles, it aligns itself to a convincing angle, then starts to sway gently as though in a breeze.

The celebrated British naturalist Alfred Russel Wallace came across *K. paralekta* in Sumatra in 1861, and he described in his book "*The Malay Archipelago*" how he had the greatest difficulty finding the butterfly once it had settled, even when he had watched it fly in and marked the spot with his eye.

Reluctant flier

Dead-leaf butterflies live in moist tropical forest, usually on hilly terrain. They are seldom seen in the open and never fly far, spending most of their time resting. Whey they do fly, as when they are disturbed, they veer off on an erratic course, the bold upperwing colors making them conspicuous. As a result they are often chased by birds, but once the butterfly has settled, closed its wings and "vanished," its pursuer is left baffled.

Kallima butterflies have the strong flight and bright upperwing coloration typical of the large family Nymphalidae.

Mimicry of dead leaves is so highly developed in the underwings of kallima butterflies that leaflike "veins" and "fungal growths" are present. Pictured is **Kallima paralekta** ***of India.***

KALLIMA BUTTERFLIES

CLASS **Insecta**

ORDER **Lepidoptera**

FAMILY **Nymphalidae**

GENUS ***Kallima*** **and** ***Kallimoides***

SPECIES **Several, including Indian leaf butterfly, *Kallima inachus* (detailed below), and African leaf butterfly, *Kallimoides rumia***

ALTERNATIVE NAME
Orange oakleaf butterfly

LENGTH
Wingspan: 4–6 in. (10–12 cm)

DISTINCTIVE FEATURES
Angular forewings; long "tail" on each hind wing; brilliant orange and blue upperside to wings; undersides colored and textured in such a way that butterfly resembles a leaf when at rest with wings closed

DIET
Adult: plant sap; flower nectar. Larva: leaves.

BREEDING
Breeding season: April–June and also after monsoon or rainy season

LIFE SPAN
Not known

HABITAT
Tropical forest

DISTRIBUTION
India east to Southeast Asia

STATUS
Not known

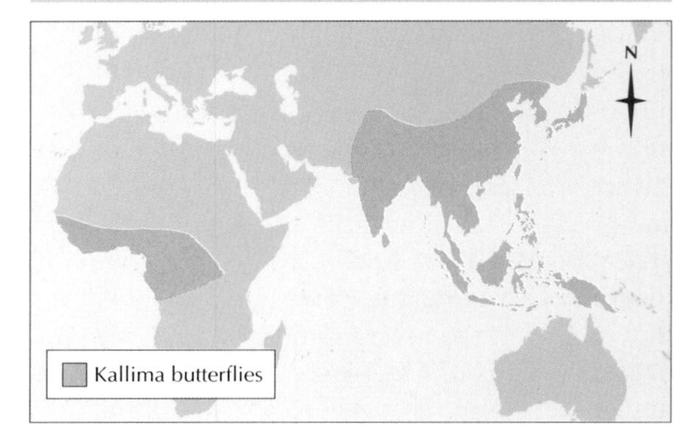

Dead-leaf butterflies often settle on sweet sap exuding from trees, on overripe fruit or on damp patches on the ground to drink. Some observers have reported seeing the butterflies hold their wings partly open, or spread and close them repeatedly, with no attempt at concealment.

Now you see it...

The Indian leaf butterfly, occasionally referred to as the orange oakleaf, is double-brooded. One breeding season runs from April to June, the other follows the rainy season. Its caterpillar feeds on flowering trees and on shrubs, including *Girardinia*, *Strobilanthus* and *Pseuderantheum*. The larva is colored a velvety black with red spines and long yellow hairs.

Once settled, the Indian leaf butterfly bears such a remarkable resemblance to a dead leaf that the species is often cited as the last word in mimicry. The species illustrates a principle that biologists call "coincident disruptive pattern." The line that extends across the fore- and hind wings and represents the midrib of the leaf only serves as camouflage when the wings are closed at rest. The same is true of the leaflike outline formed by the wings. Any change in the relative position of the wings largely destroys the illusion. However, *K. inachus* has another trick at its disposal. When the butterfly bursts into flight, there is an explosion of bright color. The overall effect is to startle and confuse any would-be predator, giving the butterfly time to flit out of immediate danger.

KANGAROO

Among Australia's best-known native animals, the kangaroos are the largest living marsupials. At up to 6 feet (1.8 m) in height, full-grown males may stand as tall as a man and weigh 200 pounds (90 kg). The head seems unnaturally small, though the ears are long; the forelimbs are short and lean by comparison with the powerful hind limbs, and the muscular tail is up to 4 feet (1.2 m) long.

The true kangaroos comprise the red kangaroo, *Macropus rufus*, and the two gray kangaroos, which once had full species status but are now usually described as races. The western gray kangaroo, *M. macropus fuliginosus*, lives in southwestern Australia, while the eastern gray, *M. m. giganteus*, occupies the full eastern third of the continent. Gray kangaroos vary in color, but are mainly gray above with whitish underparts and white on the legs and underside of the tail. The muzzle is finely haired between the nostrils. The gray lives mostly in open forest, where it browses the vegetation. The red kangaroo is found across Australia except in the extreme north, east, southwest and southeast. It is slightly larger than the gray. The male has a reddish coat, the adult female is smoky blue and the muzzle is less furry. Unlike the gray kangaroo. the red is a grazer rather than a browser and lives more in small herds, or mobs. A male kangaroo is known colloquially as a boomer, the female as a flier and the young as a joey.

Only male red kangaroos are rufous. The females (above, with young) have ashy blue coats with just a hint of red coloration.

Of the 14 species in the genus *Macropus* (literally "big feet"), only those mentioned above can strictly be called kangaroos. The others look similar: these are the eight species of wallabies and four wallaroos. There is no simple way to distinguish a kangaroo from a wallaby, except that the kangaroo is larger. As a rough rule, the kangaroo has hind feet over 10 inches (25 cm) long. Wallaroos look like short, stocky kangaroos. They include the common wallaroo or euro, *M. robustus*, a tough creature that can survive anywhere from dry bush to coastal rain forest.

The family Macropodidae, to which the genus *Macropus* belongs, also contains the rat kangaroos, tree kangaroos and pademelons, in addition to several other wallaby genera.

From a shuffle to a quick-step

When feeding, a kangaroo moves in a shuffle. First propping itself on the small forelegs and strong tail base, it swings the large hind legs forward, repeating the sequence at a leisurely pace. The famous bounding gait comes into play when speed is needed. Then only the hind limbs are needed, with the tail held almost horizontally as a balancer. Each time the leaping animal touches the ground, the thick tendons in its heels stretch and store the energy from the impact. Then, as they compress again, they catapult the feet back into the air. The result is that, once the kangaroo has built up speed, running is a highly energy-efficient method of travel. The kangaroo can top 25 miles per hour (40 km/h) over a 300-yard (90-m) stretch, although over 22 miles

The baby kangaroo, or joey, first pokes its head out of its mother's pouch at a little over 6 months old. It hops out for good at 8 months.

per hour (35 km/h) the efficiency of energy transfer is reduced. A kangaroo clears obstacles in the same way, achieving leaps of up to 26 feet (8 m). The leap usually does not lift it more than 5 feet (1.5 m) off the ground, but there are reports of large kangaroos clearing fences up to 9 feet (2.7 m) high.

Eating down the grass

Kangaroos feed mainly by night, resting during the heat of the day. The red kangaroo, with its grass-centered diet, is now a serious competitor with sheep, which are important in Australia's economy. By creating grasslands, humans have helped the kangaroo increase its numbers. In turn, the kangaroo tends to outgraze the sheep, for which the pastures were grown, not only through its increased numbers but also by its manner of feeding. Sheep have incisors (front

KANGAROOS

CLASS **Mammalia**

ORDER **Marsupialia**

FAMILY **Macropodidae**

GENUS ***Macropus***

SPECIES **Red kangaroo, *M. rufus*; western gray kangaroo, *M. macropus fuliginosus*; eastern gray kangaroo, *M. m. giganteus***

ALTERNATIVE NAME
Gray kangaroos: great gray kangaroos

WEIGHT
44–200 lb. (20–90 kg)

LENGTH
Head and body: 40–63 in. (1–1.6 m); shoulder height: 5–6 ft. (1.5–1.8 m); tail: 37–47 in. (0.95–1.2 m)

DISTINCTIVE FEATURES
Shaggy fur; long, powerful tail and hind legs; short forelegs. Red kangaroo: naked muzzle; rufous (male), smoky blue gray (female). Gray kangaroos: silvery; finely furred muzzle.

DIET
Red kangaroo: grasses. Gray kangaroos: shrubs and bark.

BREEDING
Age at first breeding: 20–42 months; breeding season: depends on rainfall; gestation period: 28–33 days; number of young: usually 1; pouch period: about 235 days; breeding interval: 7–11 months

LIFE SPAN
Up to 25 years in captivity

HABITAT
Forest, forest margins and grassy plains

DISTRIBUTION
Australia, including Tasmania

STATUS
Common

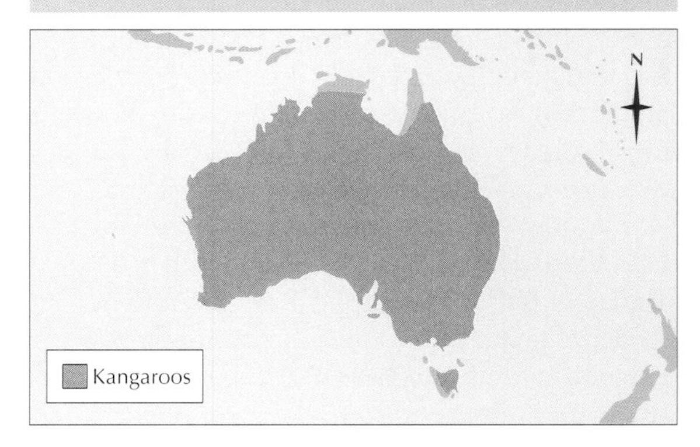

teeth) in only the lower jaw, with a dental pad in the upper jaw. Kangaroos have both lower and upper incisors and can crop grass more closely than sheep. At times kangaroos also dig out the grass roots. They can go without water for long periods, which suggests they were originally animals of desert or semidesert, although where water is supplied for sheep, kangaroos will, if not kept out, take the greater share.

Mob rule

The mobs in which both red and gray kangaroos live are loose family groups numbering two to 10 or 12 individuals. These are usually made up of females and young, joined only for breeding purposes by adults of the opposite sex. At other times the mature males live alone or busily compete with rivals for dominance of a mob. Such contests may become lively, as the males tear up grass in menacing fashion or kick and box one another. The dominant males get to mate with the most females.

An adult male red kangaroo. This species eats almost nothing but grass. It can outgraze sheep because it feeds faster and more efficiently.

Bean-sized baby

The birth act differs subtly between red and gray kangaroos. About 33 days after mating, a female red kangaroo begins to clean her pouch, holding it open with the forepaws and licking the inside. She takes up the birth position, sitting on the base of her tail with her hind legs and tail extended forward, and then licks the opening of her birth canal, or cloaca. The newborn kangaroo, just ¾ inch (1.8 cm) long, appears headfirst and grasps its mother's fur with the claws on its forefeet. Its hind legs are at this time minuscule. In 3 minutes it has dragged itself to the pouch interior and clamped its mouth over one of the four teats, which then swells to "lock" the baby in place. The process is identical for the gray except in that the female stands, with her tail straight out behind her.

The baby kangaroo is born at an early stage of development, weighing just 1/35 oz. (0.8 g). It first pokes its head out of the pouch at a little over 6 months old, then hops out for good at 8 months, by which time it weighs nearly 10 pounds (4.5 kg). It continues to be suckled for nearly 6 months after it has left the pouch and can run, thrusting its head in to grasp a teat.

Overlooking the obvious

The truth about kangaroo birth took a long time to be established. In 1629 François Pelsaert, a Dutch sea captain wrecked off southwestern Australia, was the first to discover the baby in the pouch of a female wallaby. He thought it was born in the pouch. This is what the aborigines also believed. In 1830 Alexander Collie, a ship's surgeon, investigated the birth and showed that the baby was born in the usual manner and made its way unaided into the pouch. From then on various notions were put forward: that the mother lifted the newborn baby with her forepaws or her lips and placed it in the pouch, or that the baby was budded off from the teat. It was not until 1959–1960, when the whole process of birth was filmed by researchers at Adelaide University, that the matter was set to rest.

Constant breeding

While still nursing her first joey, the female kangaroo is often preparing for the next generation. Most members of the genus *Macropus* follow a breeding strategy that is unusual among mammals. In the eastern gray and red kangaroos the female's sexual receptivity, which cycles every 45⅔ or 34⅓ days respectively, carries on for a while after conception. (In humans, by contrast, pregnancy arrests a woman's monthly periods until after the baby is born.) This means that the female kangaroo is ready to mate again within weeks of a successful mating. So, with the first baby now in the pouch, a second embryo can start to develop in the womb, and her estrous cycle is finally put on hold. It is at this stage that

An eastern gray kangaroo moving at speed. Kangaroos can top 25 miles per hour (40 km/h) over short distances. However, the real benefit of their bounding gait is that it is very energy-efficient.

matters take an unusual turn. At a stage when the second embryo consists of just 85 cells, it stops growing and remains dormant for a period of about 205 days, at which point the existing joey is about a month short of leaving the pouch. When that month is up, and the joey has gone, the embryo is born and makes its own journey to the pouch. Thereafter its growth follows the normal pattern.

This arrested-development process is known as embryonic diapause. It is a response to the special ecological conditions of the kangaroo's habitat, where local droughts can be so severe as to decimate populations. The first individuals to suffer from thirst or starvation are the joeys, so it suits the female to have another offspring "on standby" in her womb, ready to refill the empty pouch within weeks, or alternatively to develop as soon as better feeding conditions arrive. This enables populations to bounce back rapidly after a crash. At any one time, a breeding female red or eastern gray kangaroo may have a joey on the ground, another in her pouch and an embryo on hold in her womb.

Kangaroos pose a problem

Predators of the larger kangaroos once included the thylacine or Tasmanian tiger, a native marsupial carnivore that is extinct today. The dingo, *Canis familiaris*, a wild dog introduced thousands of years ago by early settlers, is still at large over much of Australia. It has been banished from the fertile sheep country of Queensland and South Australia, kept out by the vast east–west barricade known as the Dingo Fence. It is chiefly to the south of this fence that the kangaroo, with its ability to breed almost all year round, has become an agricultural pest.

Fencing in the pastures, often thousands of acres in extent, is costly, and in any case kangaroos have a trick of squeezing under fences at any weak spot, so kangaroos are often shot. In one year, on nine sheep properties totaling 1,540,000 acres (24,00 sq km), 140,000 kangaroos were shot, and it would have needed double that number of kills to clear the properties of them. Road-kills are another problem: kangaroos bound across roads at night and collide with cars, damaging vehicles and endangering human life. Many drivers now install protective "roo bars" and radiator screens, which do nothing to reduce the death toll of kangaroos. One observer recently counted 282 dead kangaroos on a 115-mile (185-km) stretch of highway in Queensland. Annual culls take place to keep numbers down, and also to supply the trade in kangaroo meat, which is tasty and lean. The skins are exported to be tanned for use as high-grade leather.

KANGAROO MOUSE

THERE ARE MANY mouse species, and many of them hop or travel in leaps. So it is not surprising that several different species that have become highly specialized in leaping should be called kangaroo mice, even when they live far from Australia and are not related to kangaroos. In Australia itself there are also several species of mice known as kangaroo mice, which some zoologists call hopping mice. In addition, on this same continent, there are 55 species of micelike animals that are not true mice and *are* related to kangaroos. These are called marsupial mice and are discussed under that title.

Australia's hopping mice

Of the 76 species of true rodents native to Australia, nine are variously known as kangaroo mice or hopping mice. They are much the same size, color and shape as the house mouse, *Mus musculus*, but their hind legs are proportionately longer. This accounts for another common name, jerboa mice, by which they are sometimes known, although they are not closely related to the true jerboas of Africa and Asia.

Many of Australia's kangaroo mice burrow into sandy soil. They are extremely difficult to dig out because of the speed with which they burrow, and once on the surface they are hard to catch because they hop along so quickly. They are vegetarians, feeding on seeds, leaves and berries. Because they do not compete for food resources, the carnivorous marsupial mice often live in the same burrows as the hopping mice. There are two to nine young in a litter, and their life span is believed to exceed three years.

Hopping around the world

In New Guinea lives another true rodent, the New Guinea kangaroo mouse or jumping mouse, *Lorentzimys nouhuysi*. It resembles the Australian hopping mice in size, color and shape, but little is known of its lifestyle.

Confusingly, there are two species in North America known as kangaroo mice, which belong to a different family from the Australasian forms. They are paler in color, but are much the same size, with long hind legs and large ears, though the tail is shorter. The head is large in relation to

Kangaroo mice occur in both North America and Australasia. They evolved in isolation to fill similar ecological niches. Pictured is* Notomys alexis *of Australia.

the body. Unlike Australian kangaroo mice, the American species have fur on the soles of their feet, which improves traction and enables them to hop more easily over loose sand.

American kangaroo mice remain in their burrows by day and come out at night to feed on seeds and insects. This helps them conserve precious body fluids. In habits they are very much like the American kangaroo rats of the genus *Dipodomys* (discussed elsewhere), particularly in their ability to do without water indefinitely, getting the moisture they need from seeds. No doubt at times they drink dew or get moisture by eating succulents, but they can do without even these.

The young of both American species are born in burrows in May or June. There are usually three or four in a litter, but there may be as many as seven, and there may be two litters a year.

Looking the part

It may seem confusing to call several entirely different animals by the same name, but there is a benefit gained from discussing them all here. They illustrate what is known as convergent evolution. By this is meant the principle that separate, unrelated animals sharing a similar way of life can evolve to look like each other. All, with perhaps one exception, the New Guinea kangaroo mouse, live under desert or semidesert conditions in which a hopping or kangaroo-like mode of travel has great advantages. First, hopping is the best way of moving over shifting soil. The jerboas show similar adaptations, and it is no accident that they are also known as kangaroo rats.

Second, food is naturally scarce in deserts, and hopping enables an animal to cover ground quickly while foraging. In the Americas the kangaroo mice have the edge over their relatives the pocket mice, genus *Perognathus*. The latter, which also live in arid habitats, tend to plow through the soil and pack their cheeks with a wide variety of seeds, many of which they will later discard as unsuitable to eat. By contrast, the kangaroo mice can be more selective, using greater mobility to forage from a wide food base.

There need be little surprise that so many mice should have specialized in leaping. Many animals show a tendency to do so, especially when young. Baby hedgehogs jump into the air when touched. They leap only a short distance, since their legs are not adapted for a life of jumping. Baby rats and mice are almost preadapted for this habit. Some go further and progress in leaps when adult. It requires only a slight change in their structure, a lengthening of the hind legs, and they are already on their way to joining the company of the kangaroo mice.

KANGAROO MICE

CLASS **Mammalia**

ORDER **Rodentia**

FAMILY (1) **Heteromyidae**

GENUS **American kangaroo mice, *Microdipodops* (2 species, detailed below)**

FAMILY (2) **Muridae**

GENUS **New Guinea kangaroo mouse, *Lorentzimys* (1 species); Australian kangaroo mice, *Notomys* (9 species)**

ALTERNATIVE NAMES
Hopping mice; jumping mice; jerboa mice

LENGTH
Head and body: 2½–3 in. (6.5–7.5 cm); tail: 2½–4 in. (6.5–10 cm)

DISTINCTIVE FEATURES
Very long hind legs; hairy soles; large head; silky fur; gray brown above, whitish below

DIET
Mainly seeds and insects

BREEDING
Age at first breeding: within first year; breeding season: all year; gestation period: not known; number of young: 2 to 7; breeding interval: as little as 4 months

LIFE SPAN
Up to 5 years

HABITAT
Scrubland and sand dunes

DISTRIBUTION
Western U.S.

STATUS
Fairly common; some subspecies vulnerable

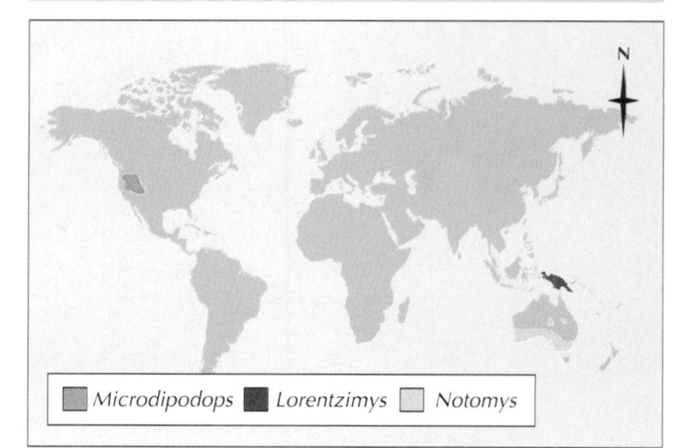

KANGAROO RAT

Kangaroo rats can survive long periods without drinking, and store food to tide them over during droughts. **Dipodomys ordii** *is shown above.*

Kangaroo rats are rodents, named for their long hind legs and tail, short forelegs and leaping gait. They are similar in these respects to jerboas and kangaroo mice and, like these animals, they live in deserts. There are 22 species, ranging in head-and-body length from 4 to 8 inches (10–20 cm), with a proportionately longer tail. The fur is yellowish to dark brown above and white underneath. A dark stripe runs along the top of the tail, which ends in a tuft of longer hairs. A gland on the upper back produces an oily secretion used by adults to scent-mark territories. Kangaroo rats live in North America west of the Missouri, from southwestern California to central Mexico.

Dry, dark and dusty

Kangaroo rats live in dry or semidry country, preferably with loose soil that they can dig easily and sparse vegetation, which favors their leaping method of locomotion. All species are nocturnal, and are reluctant to emerge even in moonlight unless compelled by hunger. Wet weather also keeps them at home.

Like many small, furred mammals, kangaroo rats take regular dust baths, without which they soon develop sores and their fur becomes matted by the oily back secretions.

Food stores

Kangaroo rats can survive long periods without drinking. They obtain their water from dew-soaked and succulent plants, but they can also live on dry food, deriving water from the breakdown of fats and carbohydrates. Their kidneys are many times more efficient than our own, and little water is needed to remove body wastes.

A variety of plant parts, including seeds, leaves, stems and fruits, are on the menu, which also includes a few insects. Kangaroo rats collect food in caches for use in times of drought, when the sparse vegetation withers They carry the food in their cheek pouches to caches and, using both forepaws simultaneously, squeeze the cheeks to empty them out. The pouches, formed from folds of skin, are lined with fur and extend back as far as the neck. Kangaroo rats near grain fields may steal enough grain to be pests.

Breeding according to the climate

Kangaroo rats breed at any time of the year, provided the climate is suitable. Some breed year-round but with few births during the winter months. Gestation lasts 4 or 5 weeks and the litters comprise one to five babies, which stay in their mother's burrow for 6 weeks. Each female may bear up to three litters in a year.

Listening for trouble

A feature of any desert-living animals such as elephant shrews, gerbils and jerboas is extremely sensitive hearing. These animals also have very large auditory bullae, the domed-shaped bones that lie under the base of the skull just beneath the ears. At one time it was thought the bullae acted as resonators that amplified the sounds

Like many desert mammals, kangaroo rats (*Dipodomys spectabilis*, above) have superb hearing.

being transmitted to the inner ears, so allowing these animals to hear very faint sounds. Today, though, scientists know this to be incorrect.

The function of the bullae and their role in the life of kangaroo rats was demonstrated by a series of experiments carried out by Douglas Webster. First, he showed that kangaroo rats are extraordinarily sensitive to sound frequencies of 1,000–3,000 cycles per second. Two mechanisms are involved. Sound waves are transmitted from the eardrum to the sense organs of the inner ear by three small bones in the middle ear. These also amplify the sound, and those of kangaroo rats amplify sound five times more than those in human ears. The large space contained in the auditory bullae was also found to increase the sensitivity of the ear. The enlarged bullae do not act as resonators but allow the eardrum to vibrate more freely. The eardrum in a normal ear is damped as a result of air pressure building up behind it in the middle ear and resisting its movements. In the kangaroo rats and other desert animals the large space inside the bullae easily absorbs any increase in pressure, and the eardrum is able to vibrate in response to much weaker vibrations.

Keen hearing is a lifesaver for the kangaroo rat. Its foraging activities take it out into the open where predators, such as owls and rattlesnakes, are active at night in search of small mammals. These predators appear to us to strike silently, but scientists have recorded faint rustling sounds given off by their actions, and these are of the frequencies to which a kangaroo rat is specially sensitive, so the rodent receives warning of an attack in just enough time to make its getaway.

KANGAROO RATS

CLASS **Mammalia**

ORDER **Rodentia**

FAMILY **Heteromyidae**

GENUS AND SPECIES **22 species, including *Dipodomys ordii*; *D. microps*; *D. stephensi*; *D. elephantinus*; *D. spectabilis*; *D. insularis*; *D. compactus*; and *D. margaritae***

WEIGHT
1¼–6⅓ oz. (35–180 g)

LENGTH
Head and body: 4–8 in. (10–20 cm); tail: 4–8½ in. (10–21.5 cm), usually longer than body

DISTINCTIVE FEATURES
Large hind feet; small forefeet; very long tail tipped with tuft; yellowish brown upperparts; white underparts

DIET
Seeds, leaves, stems and shoots; also some fruits and insects

BREEDING
Age at first breeding: 2 months; breeding season: potentially all year; gestation period: 29–36 days; number of young: 1 to 6; breeding interval: up to 3 litters per year

LIFE SPAN
Up to 9 years in captivity

HABITAT
Desert; arid zones of grasslands and scrub

DISTRIBUTION
North America, from southern Canada south to Texas, Baja California and central Mexico

STATUS
Most species common, although a few subspecies vulnerable; critically endangered: *D. insularis* and *D. margaritae*

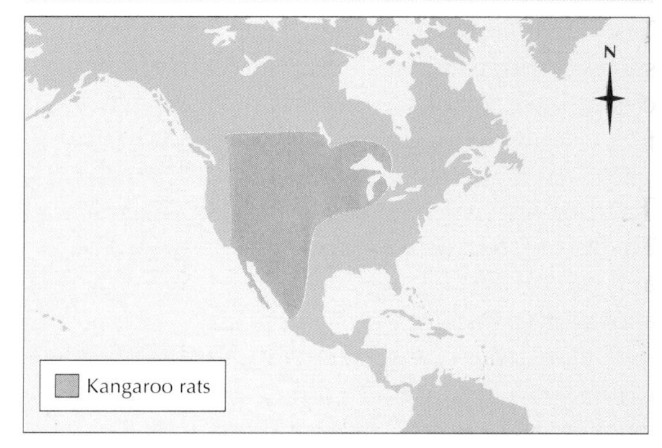

KATYDID

THE NAME OF this American group of bush crickets is said to derive from its approximation of their extremely loud call. They seem to be calling "Katy did, she did." Loudest among the crickets, these insects have also been closely studied on the subject of insect stridulation.

Two of the better known katydids are *Pterophylla camellifolia*, common in the eastern states, and *Microcentrum retinerve*, found in the southern and western states. Katydids prey mainly on other invertebrates, such as insects, but also feed on leaves, as do other bush crickets and grasshoppers. Their chief enemies are birds, and some species have effective camouflage that disguises them as leaves. In autumn they lay flattened, slate-colored eggs in two rows along a twig. These hatch the following spring, and the nymphs grow by molting at regular stages.

This katydid from the rain forest in Costa Rica is one of several species that mimic leaves. It even has brown "blemishes," just like real leaves.

One species in which the song has been closely studied is *Scudderia texensis*, the Texas bush katydid. It is common on wastelands and fallow fields, on the verges of highways and railroad embankments and wherever weeds and grasses abound. Its range covers the eastern United States. In Florida and southern Georgia it produces two generations a year, which reach adulthood in June and September. In more northerly parts of its range there probably is only one generation a year, in early July.

Songs for separate occasions

The study of katydid calls has identified several different sounds that appear to make up an insect "language." The males of the various species produce two or more different sounds even when alone, and they produce others in response to other males of their own species. The females also make their own sounds. The male Texas bush katydid makes four basic different sounds, but each can be altered according to circumstances. This can be best appreciated by following a day in the life of this katydid.

Tuning up after noon

The katydid makes no sounds in the morning and the middle of the day. In the late afternoon and evening he makes short, lisping sounds, known as the fast-pulsed song. This normally lasts less than a minute at a stretch. At twilight he begins to make a soft ticking sound that can be heard only a few feet away, interspersed with a few fast pulses. As darkness falls, he adds a slow-pulsed song. Each slow pulse is followed by a fast pulse and then a pause, and as the night wears on the fast pulse steadily lengthens, until late at night the fast pulse may last half an hour or more before the slow pulse is produced.

Dim lights and soft music

To a large extent the songs of katydids, and presumably of all bush crickets, grasshoppers and true crickets, are governed by the intensity of light. In many species a low light intensity is needed to prompt singing. In others, including the Texas bush katydid, certain sounds are produced in twilight and others in full daylight. In a few species the same sounds are produced by day and by night. The link between different intensities of light is illustrated when a katydid singing in afternoon sunlight switches to his twilight song when a cloud passes overhead. Even so, a male may change his song for a while without any change in the intensity of light. It is assumed that some change within the insect itself causes the alteration in the song.

Singing man to man

Generally, bush crickets and grasshoppers living at high population densities produce low-intensity songs that are audible to humans at a few feet only. Those in sparser populations use louder songs, which can be thought of as more of a "shout" than a "whisper." These are audible at 200–300 yards (180–275 m). The male Texas bush katydid sings to defend his territory. At first he sings to himself. If another male closes in, he advertises his presence, in effect saying "go away." If the second male moves in much closer, the resident's song becomes a threat.

Talking him in

The song plays a key role in bringing a male and female together for breeding. In some species the male calls first. A female that is ready for mating then answers him with a ticking, or occasionally lisping, call. She stays put, however, obliging the male to move over to her. This tactic is used by the Texas species: the male calls and the female ticks her answer, and the way she responds, with one, two or three ticks, tells him whether she is fully ready for mating. When he receives the maximum response, he clambers in her direction and takes flight. If he lands on a plant near her, he calls and she answers, so off he flies again. Once he has settled on the same plant, there is silence because she, sensing him land, sways gently and shakes the plant so he can find her.

In other species, the female responds to a male by coming part of the way toward him; then she begins to tick in answer to his song. He then moves to her, homing on her ticks. In a third group the female responds to the male with a series of ticks that are in effect a "come on" call. He approaches part of the way, but then changes his tune and halts. The female now goes to meet him. This differs from the behavior of other bush crickets and grasshoppers, in which the male sings and the female, who is mute, does all the traveling. There may be katydid species in which both male and female approach each other at the same time, but this has an obvious disadvantage: it is not easy to keep track of a moving sound.

Katydids are famous for their songs, which are louder and more complex than those of true crickets and grasshoppers. Both male and female sing for a variety of reasons.

Handy hearing aid

Most insects find a mate by using eyesight, as in butterflies, or scent, as in moths. Others may be drawn together at common food plants. Few use sounds to unite the sexes, and of those the katydids have the most complicated techniques for using sound. It has even been recorded that a male katydid sometimes leans to one side to lift a foreleg and hold it in the air. His hearing organs are on his forelegs, and this action seems to be a definite "straining to listen."

KATYDIDS

PHYLUM **Arthropoda**

CLASS **Insecta**

ORDER **Orthoptera**

FAMILY **Tettigoniidae**

SUBFAMILY **Phaneropterinae**

SPECIES **About 4,000**

ALTERNATIVE NAMES
Long-horned grasshopper; bush cricket

LENGTH
⅜–4 in. (1–10 cm), depending on species

DISTINCTIVE FEATURES
Resemble grasshoppers and true crickets, but with longer antennae, up to 3 times body length. Strong, elongated hind legs; 2 pairs of wings folded around back; large ovipositor (female only); generally green or brown; superb camouflage in some species, often including leaf mimicry.

DIET
Most species: insects and plant matter, such as leaves; few species: entirely predatory

BREEDING
Hemimetabolous (undergo incomplete metamorphosis). Breeding season: eggs laid in autumn; breeding interval: up to 2 generations per year.

LIFE SPAN
Up to 1 year

HABITAT
Typically bushes and other shrubby vegetation

DISTRIBUTION
Virtually worldwide; half of all species occur in Amazon River Basin

STATUS
Common

KEA

THE KEA IS AN UNUSUAL parrot, the size of a large crow, found only on South Island, New Zealand. The plumage, which has a scaly appearance, is olive green overall with streaks of orange and red on the underparts. The legs are yellow-brown and the eyes have yellow pupils that give the kea a beady gaze. The bill is nearly 2 inches (5 cm) long and extremely sharp at the tip, but it is not as curved as in other parrots. It is very strong, and the kea is reputed to be able to rip open corrugated iron roofs.

Closely related to the kea is the kaka, *Nestor meridionalis*, which is similar in both habits and appearance. The South Island variety of kaka is larger than that found on the North Island. It has predominantly brown plumage with red under the wings. Its crown is white, rather than gray, and its back is greenish. The kaka lives in dense forests, where it can be hard to spot; in kaka sanctuaries such as that on Kapiti Island, the birds are semi-tame. In common with all parrots the kea and kaka are very noisy, and their names are derived from their loud, raucous calls.

Can-opener bill

Keas are the hardiest of parrots, sometimes being found in the snow above the tree line. In winter they retreat down the mountainsides and live in the forests. The kakas keep to the lower forests. Both form flocks outside the breeding season. Keas are strong fliers, soaring and gliding gracefully from one rock or tree to the next. They are less agile on the ground, where they hop.

The kea's favorite food is the young leaves and buds of trees, but it also walks along the branches tearing off moss and lichen. Feeding is slow and deliberate, the upper part of the bill serving as a lever and the lower part as a gouge. During the southern fall (the northern spring) the kea feeds on the berries of the snow totara plant.

The tongue has a fringe of hairs reminiscent of that seen in lorikeets and lories. It is used to lick nectar from flowers or juice from succulent fruit. Unlike lorikeets, lories and many other nectar-drinking birds, keas do not pollinate the flowers on which they feed. Rather, they destroy the flowers, often chewing them to a pulp to extract the juices and spitting out the remains.

Keas extract grubs and beetles from the soil and also take carrion. Some coastal colonies prey on nestlings of the sooty shearwater, *Puffinus griseus*, digging them from their nest tunnels. Kakas feed on nectar taken up with their brush-tipped tongues, and on berries and other fruits as well as invertebrates.

Nesting under rocks

Keas occupy loose territories centered around the nest and roost. Flocks may freely enter a territory, whereupon the resident pair joins them to feed but retires to its own roost at night. Any unfamiliar kea, apparently attracted by the calls of the young, is tolerated until it comes within 25 yards (23 m) of the nest. The pair then challenges it with a cry of *kua-ua-ua-ua*, a call that is also used to challenge humans and even vehicles.

Breeding takes place year-round but mainly between July and January (spring and summer). The nest of lichens, moss, leaves, ferns and chips of rotten wood is made on the ground, under a boulder, in a crevice, in a hollow log or among the roots of a tree. There may be an entrance tunnel up to 20 feet (6 m) long, and a well-worn runway and accumulated droppings indicate that the same nests are used year after year. Two to four eggs are incubated for 3 to 4 weeks. The male roosts outside the nest, and the female joins him when the chicks are 8 weeks old and filling most of the nest. At first these are helpless, and

Keas live in a very unusual habitat for a parrot: open pasture and scrub in high mountains. In summer, they often venture above the snow line.

Keas are bold, playful and inquisitive birds accustomed to human presence. They damage property and shred tent fabric or unattended clothes at camping sites.

the female feeds them by pushing food into their mouths. Soon she loses interest, and the male takes over feeding. Young males become independent 4 weeks after fledging and disperse, but the father continues feeding the young females for another few weeks.

Kakas nest in hollow trees, the female laying two to five eggs on a layer of powdered wood in a cavity in a tree.

Making mischief

In a country renowned for its shy wildlife keas stand out as bold, inquisitive, intelligent and playful birds. On sites such as farms and ski resorts, where the birds are habituated to human activity, they regularly cause damage to property, tampering with vehicles and machinery and shredding tent fabric or unattended clothing. Although main offenders have to be rounded up and kept in captivity, the keas' lively antics in general have become an unofficial but highly endearing tourist attraction.

Keas have another, less welcome, reputation. It was once thought that they killed sheep, and so they were persecuted. In 1886 the government offered a bounty for each kea bill. In the period 1943–1946 alone, 6,819 keas were killed. It is now known that attacks on living sheep are rare and involve only animals that are already diseased or injured. The species, whose range falls within several national parks, received partial protection in 1970, upgraded to full protection in 1986. This, along with research into breeding biology, is hoped to reverse the population decline.

KEA

CLASS **Aves**

ORDER **Psittaciformes**

FAMILY **Psittacidae**

GENUS AND SPECIES ***Nestor notabilis***

WEIGHT
Average 2 lb. (920 g)

LENGTH
Head to tail: about 19 in. (48 cm); male larger than female

DISTINCTIVE FEATURES
Male: hooked bill, with long upper mandible; brown head and underparts; bronze-green back and wings, with black scaling; reddish underwing, rump and uppertail coverts; tail bronze-green above, yellow below. Female: shorter, less hooked bill. Juvenile: yellowish crown.

DIET
Mainly leaves, buds, flowers, berries and fruits; some insects, carrion and scraps; also seabird chicks (coastal populations only)

BREEDING
Age at first breeding: 1 year; number of eggs: 2 to 4; incubation period: 3–4 weeks; fledging period: about 13 weeks; breeding interval: variable

LIFE SPAN
Not known

HABITAT
Summer: areas of scrub and pasture in high mountains, above tree line. Winter: wooded valleys and forests.

DISTRIBUTION
Mountains of South Island, New Zealand

STATUS
Near threatened; estimated population: 1,000 to 5,000

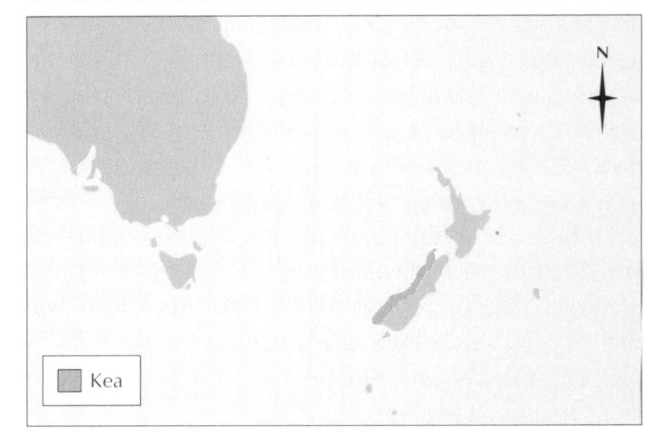

KESTREL

A Eurasian kestrel feeding on a pigeon. Kestrels take a wide range of prey, including butterflies and earthworms.

THE KESTRELS ARE SMALL falcons, found on all the continents except Antarctica. Some are noted for hovering on gently fanning wings while searching the ground below for prey. Thirteen species bear the name kestrel; in all the plumage is chestnut and gray with black spots in the male and pale reddish brown with black streaks and bars in the female.

The Eurasian kestrel, *Falco tinnunculus*, has many races throughout Europe, Asia and Africa. It is about 13–14 inches (33–36 cm) long, the male slightly larger than the female. The mature male has a gray crown, whereas the female and young have brown crowns. A young male obtains his gray crown at two years. The American species, *F. sparverius*, is one of the smallest species in the genus, at 7½–8¼ inches (19–21 cm) long. In this species the male is distinctly more reddish in color than the female. The American kestrel is found over the length of the continents, from the northern tree line to Tierra del Fuego, but it shuns heavily forested regions, such as Amazonia.

There are several species on the islands of the Indian and western Pacific Oceans. Rarest of them all is the Mauritius kestrel, *F. punctatus*. Conservaion measures have boosted its population from just four wild birds in 1974 to about 65 pairs in the wild in 1994. The population is expected to rise to 500 to 600 birds. The largest kestrels are two from Africa: the greater kestrel, *F. rupicoloides*, and fox kestrel, *F. alopex*. Neither hovers except very rarely. They take insects, reptiles and small mammals on the ground.

Hovering for a living

The Eurasian kestrel nests and roosts in open woodland but hunts over open country. It also takes up residence in the niches of tall buildings in major cities, preying mainly on house sparrows and starlings. Besides hovering, it takes up position on a perch on a bush, wall or building, or on a post or telegraph wire, and from such vantage points it drops to the ground to snatch prey. When hovering, with head into the wind and tail bent down and fanned, the kestrel may glide on the slant to take up a new hovering station, or it may drop to the ground and then fly up again to a new position. In straight flight it alternates glides with a few quick wingbeats. The call is a loud *kee-kee-kee*.

The Eurasian kestrel feeds mainly on mice and voles, as well as small birds, insects and earthworms, and this is typical of all kestrels except the two large African species. Insect prey

The American kestrel (male, above) is one of the smallest kestrels. It is found from the northern tree line in Canada all the way to Tierra del Fuego at the southernmost tip of South America.

comprises large beetles, moths, grasshoppers and the like. The overall balance of the diet depends much on the season and on local abundance. For example, it is not unusual for a kestrel to spend an hour or more hovering over one field, dropping to earth every now and then, eating nothing but butterflies and moths. At another time a kestrel may spend long periods dropping from a post and back again, taking nothing but earthworms. Kestrels also take carrion, such as large bird carcasses, and both the Eurasian kestrel and the American kestrel have been seen taking meat scraps and bread from bird feeders. They also rob other birds of prey. One kestrel was seen flying up to a barn owl carrying a vole; turning on its back below the owl, the kestrel seized the vole in its talons and then flew away.

Aerial courtship

Courtship, which takes place in late March or early April, consists of aerial displays by the male flying in circles over the perched female.

EURASIAN KESTREL

CLASS **Aves**

ORDER **Falconiformes**

FAMILY **Falconidae**

GENUS AND SPECIES ***Falco tinnunculus***

ALTERNATIVE NAME
Windhover (archaic)

WEIGHT
Male: 5½–7½ oz. (155–215 g). Female: 6⅘–8⅞ oz. (195–250 g).

LENGTH
Head to tail: 12½–13¾ in. (32–35 cm)

DISTINCTIVE FEATURES
Male: blue gray crown, nape and tail; chestnut upperparts with black markings; buff underparts with dark streaks; black tail band. Female: brown crown and tail.

DIET
Small mammals, small birds, invertebrates and lizards

BREEDING
Age at first breeding: 1 year; breeding season: eggs laid March–April; number of eggs: 3 to 6; incubation period: 27–29 days; fledging period: 27–32 days; breeding interval: 1 year

LIFE SPAN
Up to 16 years

HABITAT
Most available habitats; absent only from tundra and isolated taiga

DISTRIBUTION
Europe; large parts of Asia (absent from most of Siberia and Indian subcontinent); North and sub-Saharan Africa; southern and eastern Arabia

STATUS
Common

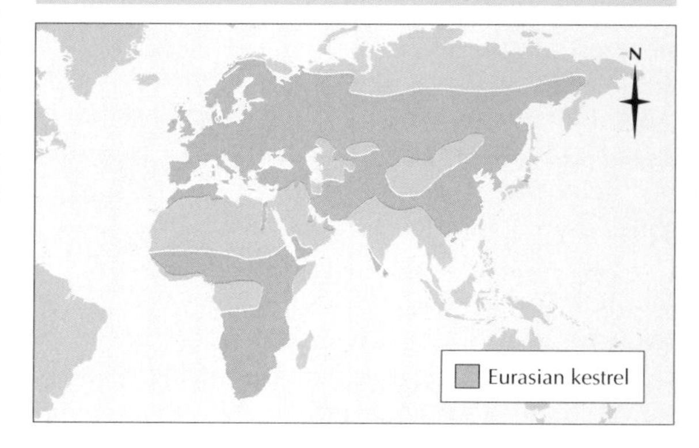

A female Eurasian kestrel with a brood of 7-day-old chicks. She is feeding them a young starling, brought to the nest by her mate.

Throughout his displays he flies with three or four wingbeats followed by a glide, repeated as if part of a ritual, and calling *kee-kee-kee* all the time. Every so often he flies at the female, not stooping at her but merely buzzing her, pulling out of his dive at the last moment to fly up and circle her again. Sometimes she flies up, and he continues the maneuvers over and around her.

Kestrels make no nest, so the three to six eggs, white with blotches of dark reddish brown, are laid at intervals of two days in the abandoned nest of a large bird such as a crow, in a crevice in a building, on a ledge on a cliff or in a hollow tree. Incubation is mainly by the female, the male bringing her food, although he does occasionally brood. The eggs hatch in a month, and again it is the hen that broods the nestlings while her mate fetches food. The chicks fledge at about a month, but are fed for a further period after fledging. When they finally disperse, the nest is littered with pellets of indigestible prey remains that have been regurgitated by the young.

Defensive tactics

A young kestrel reacts defensively by rolling onto its back to present its sharp talons to the intruder. When the intrusion is a human hand, the talons take firm hold of it, and the bird clings on obstinately, later bringing its bill into use. The juvenile's defense reaction, which is shared by other birds of prey including owls, is probably enough to keep most would-be predators at bay. Accidents do happen, though: one kestrel was seen to stoop on a vole just as a weasel was about to do the same. The weasel killed the kestrel.

Keen-sighted killer

The kestrel's archaic country name of windhover is apt, as the bird may hover at heights of just a few feet up to 100 feet (30 m) or more, turning its head from side to side as it scans the ground below. At the same time, it is not unaware of movement to the side. This may be illustrated by a kestrel that was seen hovering over a field, when it suddenly glided away to the right to the top of a tall oak tree 200 yards (185 m) away. A birdwatcher positioned at a similar distance from the oak could see nothing with his naked eye on the foliage of the oak to attract the kestrel. When he brought his binoculars up, however, he could see that the kestrel had taken a small white butterfly in its bill.

Recent research has shown that the kestrel's eyes can detect ultraviolet light. This is useful for a bird that preys on voles, which leave regular traces of urine to mark out their runways in the ground cover. The urine reflects ultraviolet light, enabling the kestrel to home in on the voles.

KILLDEER

A COMMON PLOVER OF the United States, the killdeer breeds on inland pastures and meadows. The adult is 10 inches (25 cm) long, rather larger than the related ringed plovers but similar in plumage. The upperparts are bright chestnut and the underparts white. The forehead and breast have bold white bands. There are two bands on the breast compared with the single band on the ringed plover and the semipalmated plover. The long legs are straw-colored, and the bill is black.

The killdeer breeds from Canada south to Mexico, including the Caribbean islands. The northern population migrates south for the winter, and occasional stragglers find their way across the Atlantic to western Europe. The species is unique among shorebirds in being occasionally swept northeast up the Atlantic coast of North America by late fall or winter storms. The handful of birds that reach western Europe probably arrive that way.

The killdeer is named for its very distinctive two-note call,* kill-dee*, which it often utters when taking off.

Hard to see, easy to hear

Outside the breeding season killdeer live in flocks, inland or on shores. Despite their striking plumage, the flocks can be difficult to see, and because they are quite tame one can sometimes walk almost up to a flock before noticing them. The black bands are an example of a disruptive pattern, breaking up the outline of the body and making it harder to see. By contrast, killdeer are very noisy, as indicated by the second part of the species' scientific name, *Charadrius vociferus*. The common name derives from their call note of *kill-dee*. When disturbed, killdeer fly very rapidly, not far off the ground, and they can also run fast.

Beneficial birds

When feeding in a flock, killdeer scatter rather than form a compact bunch. Feeding by day and night, they search among grass or low plants for insects and other small prey, and also follow the plow. In one study, two-thirds of the animals found in the stomachs of killdeer were insects, of which one-third were beetles. Others included grasshoppers, flies, ants, weevils and bugs, as well as caterpillars. There were also centipedes, ticks, earthworms and snails.

Many of the killdeer's favorite prey items are pests, such as weevils, wireworms and caterpillars. Killdeer are key predators of the cotton boll weevil; one bird had 380 weevils in its stomach. The birds have also earned a good name through eating mosquito larvae and cattle ticks.

Nesting on pebbles

The male killdeer performs a spectacular aerial display. He flies back and forth or climbs in a spiral, sometimes hovering, for up to an hour at a time. While in the air, he gives voice to a continuous, musical trill. On the ground the female is courted with a display in which the male lowers his spread wings to the ground and fans his tail over his back.

The nest is no more than a depression in the ground lined with a few pebbles and other bits and pieces. Killdeer like to nest on stony ground, in meadows, pastures or cultivated fields where the ground is bare, and usually fairly near water. Unusual nesting sites include the ballast on railway tracks, a tar-and-gravel roof and a garbage heap of old bottles and cans. The eggs are three to five in number, pale brown with brown or black irregular marks. Both parents incubate the eggs for 24–28 days, and both take turns to guard the young birds, which leave the nest shortly after they have hatched.

KILLDEER

CLASS **Aves**

ORDER **Charadriiformes**

FAMILY **Charadriidae**

GENUS AND SPECIES ***Charadrius vociferus***

WEIGHT
2½–3¼ oz. (72–93 g)

LENGTH
Head to tail: 9–10¼ in. (23–26 cm); wingspan: 23¼–24¼ in. (59–63 cm)

DISTINCTIVE FEATURES
White underparts with double black breast bands; chestnut-brown upperparts; black eye with red eye-ring; black bill; rufous rump and center to tail, with bold black-and-white surround when spread

DIET
Invertebrates such as beetles, bugs, flies, grasshoppers, caterpillars, ants and weevils

BREEDING
Age at first breeding: 1 year; breeding season: March–July; number of eggs: 3 to 5; incubation: 24–28 days; fledging period: about 31 days; breeding interval: 1 year

LIFE SPAN
Up to 6 years

HABITAT
Summer: open fields and meadows with low vegetation; fledged young move to wetter habitats in valleys or on flood plains. Winter: usually coastal habitats.

DISTRIBUTION
Breeds from Alaska and central Canada south to central Mexico and Caribbean; northern birds are migratory, wintering in northern South America

STATUS
Common

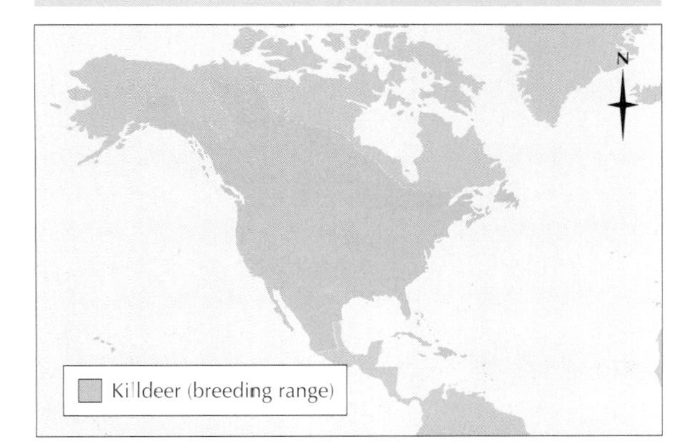

If a predator comes near a killdeer's nest, the bird pretends to have a broken wing and flutters weakly away, encouraging the enemy to follow. When the intruder is at a safe distance, the killdeer suddenly takes off.

Feigning injury

The eggs and chicks of killdeer are superbly camouflaged, as is the rule among ground-nesting birds. If confronted with an intruder, the adults take the extra precaution of luring it away from the nest site with a distraction display. Such behavior is also common among ground-nesters, especially shorebirds. The most usual display is that known as injury feigning, in which the parent gives a passable imitation of a bird with a broken wing and lures a predator, such as a fox, coyote or raccoon, away from its brood and then flies up into the air just before it is attacked.

Any humans wandering near a killdeer's nest are treated to a similar display. One or other parent flutters to the ground with one wing held over its back, the other beating against the ground and the tail fanned. The effect of such theatrical antics is heightened by the exposure of wing and tail plumage. If the bird is approached, it normally runs a short distance and repeats the performance. It continues to do so until the intruder is about 100 yards (90 m) from the nest, at which point the killdeer flies away to safety, its ruse successfully completed.

When confronted with grazing animals, such as cattle and sheep, which are hazardous because they may trample the brood, a killdeer reacts differently. It stands over the nest or young with wings spread and quivering, calling vigorously. That this warning display is effective was proven on one occasion when a flock of goats was seen running in a direction that would lead it over a killdeer's nest. An adult at once began a warning display at the nest, and the flock parted down the middle, running past it to either side.

KILLER WHALE

THE KILLER WHALE HAS A doubly unsuitable name. It is, strictly speaking, a dolphin, the largest of the family Delphinidae, whose 30 or so species include the closely related false killer whale, *Pseudorca crassidens*, and pilot whales, as well as the much more familiar beak-snouted species. Moreover, despite its wide reputation for great ferocity, the "killer" is no more murderous than any other large, strong predator. The orca, as it is also known, is a powerful, highly intelligent mammal adapted to living in close-knit family groups called pods. It hunts collectively.

The female grows to a maximum length of about 28 feet (8.5 m), but usually stops at around 15 feet (4.5 m). A mature male may be as long as 32 feet (10 m), but again is typically shorter. This marked sexual dimorphism (difference between the sexes) is rare among whales, though it is seen also in the sperm whale, *Physeter catodon*.

In addition to its impressive size, the killer whale's coloration is distinctive. Both sexes are black above and white below. Occasionally the white is yellowish. The chin is white, and there is a white oval streak just above and behind the eye. There is a small, pale patch just behind the dorsal fin that varies in shape and hue from one animal to the next. The white on the underside sweeps up toward the tail, and the flanks are white between the dorsal fin and tail. The flippers, which are broad and rounded, are black all over, whereas the tail flukes are white below. The dorsal fin is conspicuously straight and tall in the male, usually about 2 feet (60 cm) high but growing to 6 feet (1.8 m) in a mature individual. Breaking the water in a graceful arc when killer whale pods are breathing at the surface, the tall fin is a useful guide to identification. The female has the sickle-shaped dorsal fin more expected of a dolphin. An old male may also have very long flippers, up to one-fifth the animal's total length; that proportion in young males or adult females reduces to only one-ninth.

Killer whales are found in all seas but are specially numerous in the Arctic and Antarctic, where there is abundant food to satisfy their large appetite. They favor coastal waters, where prey stocks are greatest, and can often be seen in estuaries and bays. Some populations are migratory, perhaps following the movements of prey. One such is the North Atlantic population, which heads farther north during the warmer months into food-rich polar waters.

Living in groups

The pods in which killer whales live number from two to several dozen animals of either sex and any age group. A typical pod comprises four or five females and calves led by a large, mature alpha male and his mate. The composition of larger pods is such that calves are always the most numerous individuals, followed by young females, young males, and breeding adults. Pod

A female killer whale at full breach. Also known as orcas, killer whales are the largest species of dolphin.

KILLER WHALE

CLASS **Mammalia**

ORDER **Cetacea**

FAMILY **Delphinidae**

GENUS AND SPECIES ***Orcinus orca***

ALTERNATIVE NAMES
Orca; grampus; great killer whale; blackfish; sea wolf

WEIGHT
2⅓–2½ tons (2.5–2.9 tonnes)

LENGTH
Male: up to 32 ft. (10 m); female smaller

DISTINCTIVE FEATURES
Black upper body; white or cream chin, belly and underside of tail flukes; white flash behind eye; grayish patch behind dorsal fin. Male: stocky; steeple-like dorsal fin. Female: less stocky; sickle-shaped fin.

DIET
Mainly fish and squid; also seabirds, other dolphins, turtles, seals and sea lions; very occasionally large baleen whales

BREEDING
Age at first breeding: 15 years (male), 12 years (female); breeding season: all year; gestation period: average 515 days; number of young: 1; breeding interval: 3–8 years

LIFE SPAN
Up to 50–90 years

HABITAT
Usually coastal waters where prey is abundant, but known to dive to depths as great as 1,100 yd. (1,000 m)

DISTRIBUTION
Oceans worldwide, especially in cooler waters toward poles

STATUS
Generally uncommon, but not threatened

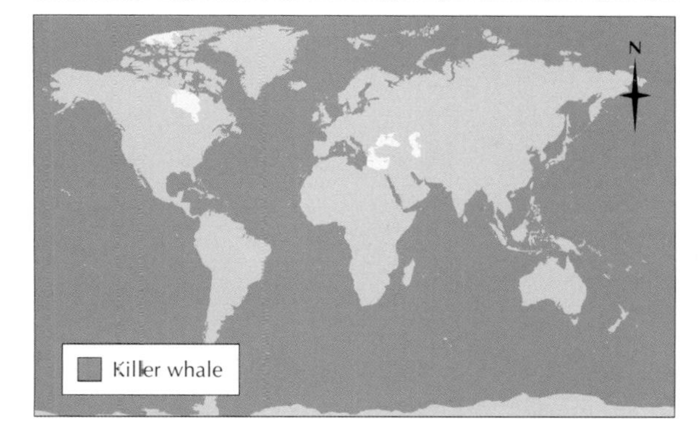

Killer whales live in family groups called pods. A typical pod is made up of four or five females and calves led by a large, mature male and his mate.

members keep in touch constantly, using an apparently complex language of clicks, whistles, screams and pulsed calls. They not only collaborate when hunting, but also come to the aid of sick or injured family members. Highly inquisitive, they take a close interest in anything likely to be edible. In the Antarctic they loiter around whaling vessels, an easy source of food. When on the move, the pod cruises at around 7 knots (13 km/h), the speed soaring easily to 29–33 knots (33–45 km/h) when required.

Warm-blooded prey

The killer whale probably earned its name from the fact that it is the oceans' largest predator of mammals, but this greatly understates its broad diet. Its staple is fish, although the killer whale may also prey on squid, penguins, seals, other dolphins and whales—even the blue whale.

Several pods may join up to form hunting packs of 40 or more. When attacking a large whale, they work as a team. First, one or two seize the tail flukes to slow the whale down and stop it from thrashing about. Others then go for the head and try to bite the lips. As the whale gradually becomes exhausted, its tongue lolls from its mouth, a rich delicacy to be seized and removed by the killers. It is now all over for the whale, and the hunters take their fill, usually from around the head region. Adding to their occasional attacks on full-grown whales, killer whales have earned the hate of whalers because they often take the tongues from whales that have been harpooned and are lying alongside the factory vessel awaiting processing.

Seal killers

More usually, killer whales prey on seals and porpoises, and there are a number of records of complete seals found in a killer's stomach. A total of 13 porpoises and 14 seals were once found in the stomach of a killer whale, while another held 32 full-grown seals. Off the Pribilof Islands in the Bering Sea, killer whales are often seen lying in ambush for fur seal pups swimming out into the open sea for the first time. Along the coast of Argentina killer whales have been filmed rushing up into the surf zone of a beach to snatch sea lions from the shore.

In the Antarctic, penguins form an important part of the diet, tempting pods to head south into the pack ice, often taking easy routes in the wake of ice-breakers. Here, too, seals are a rich food source, specially when pups or nursing mothers are confined to ice floes. Scientists based on the Antarctic Peninsula have witnessed killer whales swimming beneath ice floes and then surging up under the floe to tip it over or smash it, toppling the seals into the water and into the waiting jaws of the killers. Another technique used by a pod was to breach in unison and send a wave washing over the floe to unseat the seals.

A single calf

Breeding can occur at any time of the year, although mating peaks in May–July in the Northern Hemisphere. The female gives birth to a single calf after 15–16 months. The newborn is about 6½–13 feet (2–4 m) long and weighs about 400 pounds (180 kg). A calf is weaned at about 14–18 months old and is not likely to breed successfully until 12–18 years old.

Humans the only enemy

Killer whales face no natural predators. Despite a few reports of attacks on divers, there is no substantiated evidence that the killers themselves are a menace to humans unless provoked. Human activities have always posed a far greater threat to killer whales. During the 20th century until the mid-1980s Norway, Greenland and Japan hunted small quotas commercially for meat and oil. A more insidious danger comes from marine pollution. Mercury, PCBs and other toxic substances are passed through the food chain and rapidly build up in the bodies of cetaceans, particularly those, like killer whales, that hunt in pollution-prone coastal waters.

Tamed killer whales are crowd-pleasers at oceanaria, where they perform tricks, but there is mounting opposition to such spectacles, with even Hollywood taking a conservationist stance in the *Free Willy* movies. Eco-friendly tourism now caters to people who are prepared to watch this magnificent marine mammal in the wild.

Killer whales favor coastal waters such as bays and estuaries, where most prey is to be found. Their tall dorsal fins are a useful guide to identification.

KING CRAB

THE KING CRAB, SOMETIMES called the horseshoe crab, is not a true crab and is more nearly related to spiders. It barely even resembles a crab, looking more like a creature time forgot, and it is often described as a living fossil. Seen from above on a sandy shore, the king crab appears to be made up of a rounded, brown or dark olive-green dome hinged to a hard and roughly triangular abdomen, tipped with a long, movable tail-spike. The whole animal may measure up to 2 feet (60 cm) long.

Viewed from beneath, the dome, or carapace, is seen to shelter a series of pairs of jointed limbs. Behind the first short pair are four pairs of longer limbs. All these limbs end in little pincers except for the second pair (or second and third pairs, according to species) in the adult male. Next in the series is one more pair of legs, which lack pincers. Instead, their penultimate joints support spines that the animal uses to get a grip on the sand. Finally on the underside of the carapace is a pair of small structures, known as chilaria, of uncertain function. On the abdomen is a cover, or operculum, with the paired genital openings on its undersurface. Behind it are five pairs of flaps, or gill books, so called because each one is made of up to 200 thin leaflets. These are the gills, which extract oxygen from the water.

Of the five species, *Limulus polyphemus* lives on or near the shore in sounds, bays and estuaries down the Atlantic coast of North America from the Gulf of Maine to the Gulf of Mexico. It is particularly common in Long Island Sound and the mouth of the Delaware River. In places it is so common it has been caught in large numbers, ground up and used either as a fertilizer or as chicken feed. The other four species live along the coasts of Asia from Japan and Korea south to the Philippines, the Malay Archipelago and the Bay of Bengal.

Living sand plow

The tail-spike, or telson, of the king crab, a harmless and far from agile animal, is not the weapon it seems but is used as a lever when the crab is plowing through sand or mud. The crab also uses the telson to right itself on the rare occasions when it has been overturned by the waves or when it lands upside down after swimming.

On the beach the king crab plows through the sand by shoving with the last pair of legs; the forward limb pairs help raise the animal's front. Underwater it travels over the seabed in a more lively, bobbing gait, with the hind legs pushing the body off the seabed and the carapace continually threatening to nose-dive back again. The crab is also able to swim upside down in a leisurely manner by flapping its gill books, an action also important in circulating water among the leaflets. Much of the time a king crab rests partly or completely buried. On the top of the carapace there are two compound eyes, set to the sides, with a second pair of simple eyes farther forward, near the center.

Chews with its legs

Although it may eat seaweed, the king crab eats mainly mollusks and worms, and it is sometimes a significant predator on coastal clam beds. The mouth is surrounded by the legs, and does not have jaws. Instead, the spiny basal joints of the legs are gnashed together to chew food.

Comes ashore to breed

The American king crab breeds early in summer, when the moon is full and the tides are at their deepest. Each female creeps up the beach with a male, slightly smaller than her, clinging to her abdomen. She scoops a depression in the sand

King crabs are neither crabs nor crustaceans and belong to an ancient class of their own, Merostomata. King crabs very similar to those living today have existed for almost 200 million years.

American king crabs on a spawning beach. King crabs come ashore to breed on high tides in late spring and early summer.

near the limit of high tide and deposits in it 200 to 300 eggs, each 5 millimeters in diameter and covered with a thick membrane. As she lays the eggs, the male fertilizes them. She then scrapes sand over the clutch, and the two part company. After a month or so a short-tailed larva, about 1 millimeter long, hatches out. Shedding its shell more than 16 times as it grows, the king crab reaches sexual maturity at 9–12 years of age.

A class of their own

Not only are king crabs unrelated to true crabs, they are not even crustaceans. They form a class of their own among the arthropods. Their occurrence in two isolated areas so far apart suggests that they are relicts of a once more widespread group. Fossils in Bavaria, Germany, show that king crabs very similar to those living today have existed for almost 200 million years. The ancient order Xiphosura, to which the king crabs belong, arose about 350 million years ago, but all its other members are long extinct.

Today the king crab is the focus of a major biomedical industry. Its blood is highly sensitive to viruses and is used to test the purity of human blood samples. Also, study of its central nervous system has taught scientists much about other life-forms. In Asia, unscrupulous collection for bioresearch may soon threaten local populations.

KING CRABS

PHYLUM **Arthropoda**

CLASS **Merostomata**

ORDER **Xiphosura**

GENUS ***Limulus, Tachypleus, Carcinoscorpio***

SPECIES **4, including American king crab, *Limulus polyphemus* (detailed below)**

ALTERNATIVE NAMES
Horseshoe crab; horse-hoof crab

LENGTH
Length including telson (tail-spike): up to 2 ft. (60 cm); width across carapace (shell): up to 14 in. (35 cm)

DISTINCTIVE FEATURES
Brown or olive-green domelike shell, hinged to short, spine-edged abdomen; 6 pairs of limbs; long, mobile tail-spike

DIET
Mollusks and worms living in sandy seabed

BREEDING
Age at first breeding: 9–12 years (female a little older than male); breeding season: late spring–early summer, on beaches at high tide; number of eggs: up to 20,000; hatching time: about 30 days

LIFE SPAN
Not known

HABITAT
Tidal and subtidal zones of gentle sandy or muddy bays and beaches

DISTRIBUTION
Atlantic coast from Gulf of Maine south to Gulf of Mexico; some records from New Brunswick, Novia Scotia and Sable Island

STATUS
Not in any evident danger

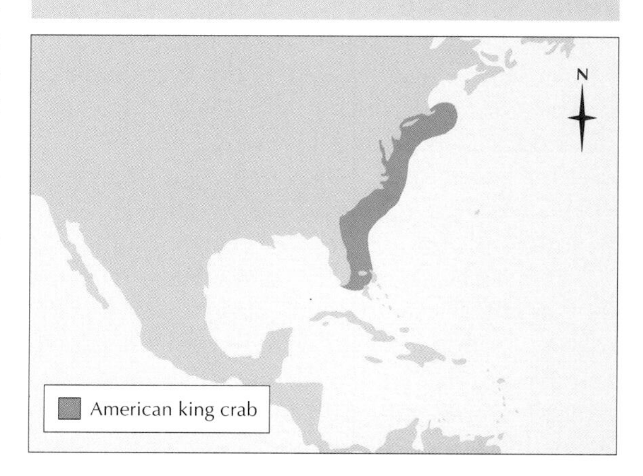

KINGFISHER

There are 93 species of kingfishers, found mainly in the Tropics. Their defining characteristics are exemplified in the common kingfisher, *Alcedo atthis*, of Europe and Asia. A stocky bird, it has a long, pointed bill, a relatively short tail and brilliantly colored plumage. Another familiar member of the family is the kookaburra, *Dacelo gigas*, of Australia; it is treated under a separate heading.

The common kingfisher is found in Europe and parts of North Africa and across Asia to the Solomon Islands and Japan. It is 6¾–7¾ inches (17–19.5 cm) in length with a 1½-inch (4-cm), daggerlike bill. The upperparts are an iridescent blue or green, the underparts are a warm chestnut, the legs are red, and there are patches of white on the neck. The pied kingfisher, *Ceryle rudis*, of sub-Saharan Africa and southwestern Asia, is relatively lacking in color but makes up for that with its bold black-and-white plumage. Like many kingfishers, it has a crest. The Amazon kingfisher, *Chloroceryle amazona*, is also crested. It has brilliant green upperparts and white underparts, with a chestnut breast in the male. The green kingfisher, *C. americana*, which ranges north into the southern United States, is very similar in appearance. In some of those species in which the sexes differ in plumage, the female is the more brilliant. On the other side of the Pacific the yellow sacred kingfisher, *Halcyon sancta*, is found in many parts of Australia and is the only kingfisher in New Zealand.

A blur of color

Kingfishers are usually seen as little more than a blur of color as they fly low over the water on whirring wings to vanish into waterside undergrowth. When glimpsed on a branch or rock on the bank, their true colors can be appreciated. That said, some kingfishers rarely, if ever, go near water. Even the common kingfisher, which is associated so much with streams and rivers, sometimes nests some distance from water. Such is the case in the Algarve region of Portugal, where kingfishers may nest several miles from the coast, in open woodland or on farmland.

In years gone by, when thousands of exotic birds were being slaughtered and their carcasses and feathers sent to Europe and North America as decorations, the dazzling kingfisher did not escape persecution. Its feathers were used on hats, and stuffed kingfishers in glass cases were a common household ornament. In later years kingfishers were shot because they were alleged

A malachite kingfisher,* Alcedo cristata*, of sub-Saharan Africa. Typical features of the kingfishers are long, daggerlike bills and brilliant plumage.

The Amazon kingfisher is a large species found in Central and South America. Kingfishers have to maneuver their prey into a suitable position before they can swallow it.

to eat enough young fish to damage breeding stocks. Today, the pollution of freshwater rivers and streams threatens their well-being. Hard winters, too, have a severe effect on populations.

Catch of the day

Nearly all kingfishers hunt by waiting on a perch, darting out to catch prey then carrying the prize back to the perch. The common kingfisher flies out, hovers momentarily just over the water and then dives in. Having caught a small fish or water insect, it uses its wings to "fly" through the water and then up into the air without pausing. It bludgeons larger prey against the perch to subdue it, and may toss and catch it again in a suitable position for swallowing.

Common kingfishers take mainly fish, such as minnows, sticklebacks and gudgeon, as well as small perch and small trout. These last two account for much of the species' former persecution. They also feed on water beetles, dragonfly larvae, small frogs, tadpoles and pond snails.

Fishing over land

Most kingfisher species, however, take mainly land animals, although they hunt from a perch like the common kingfisher. They dart down from the perch like shrikes, or catch passing insects like flycatchers. The racquet-tailed kingfisher, *Tanysiptera galatea*, which ranges from the Molucca Islands to northeastern Australia, hunts for lizards, centipedes and insects in the leaf litter of humid forests. It swoops on them, occasionally driving its bill into the soft soil. The stork-billed kingfisher, *Pelargopsis capensis*, of India, catches fish as well as frogs, lizards, crabs and insects in

COMMON KINGFISHER

CLASS **Aves**

ORDER **Coraciiformes**

FAMILY **Alcedinidae**

GENUS AND SPECIES ***Alcedo atthis***

WEIGHT
1⅕–1⅔ oz. (34–46 g)

LENGTH
Head to tail: 6¾–7¾ in. (17–19.5 cm)

DISTINCTIVE FEATURES
Compact body with relatively large head; long, daggerlike bill; short tail; brilliant blue and greenish blue upperparts; orange or chestnut underparts; white chin; reddish orange legs and feet

DIET
Mainly freshwater fish; some aquatic insects, tadpoles, small frogs and pond snails; also marine fish (coastal populations only)

BREEDING
Age at first breeding: 1 year; breeding season: eggs laid April–May (temperate Eurasia); number of eggs: usually 6 or 7; incubation period: usually 19–21 days; fledging period: 23–27 days; breeding interval: up to 3 broods per year

LIFE SPAN
Up to 15 years

HABITAT
Generally close to clear fresh water with easy access to suitable nesting banks; in some regions also in open woodland or cultivated land far from water

DISTRIBUTION
Europe, parts of North Africa, Middle East, Central and Southeast Asia, Sunda Islands, Sulawesi and some Polynesian archipelagos

STATUS
Locally common

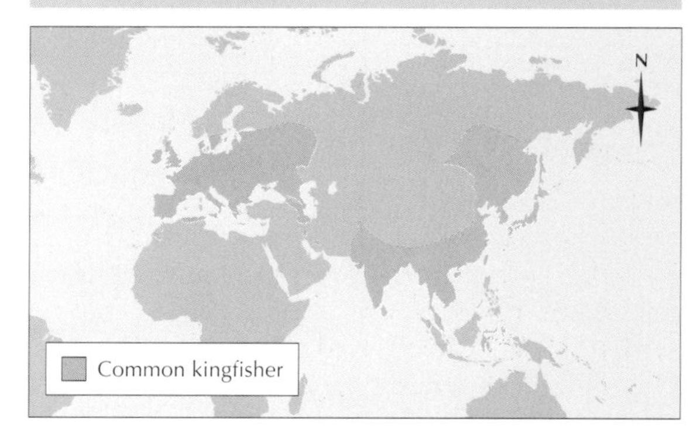

its long, scarlet bill. It also robs other birds' nests, thieving nestlings even from nests in holes in trees. True to its kind, however, it returns to its perch, often on a telegraph pole or tree, to swallow its prey. An exception to the rule is the shoe-billed kingfisher, *Clytoceyx rex*, of the forests of New Guinea, which digs for earthworms with its unusual, flattened bill.

Hole nesting

Kingfishers nest in holes, those that hunt fish usually using tunnels in banks near water while the more land-based species nest in tree holes or abandoned termite nests. The striped kingfisher, *Halcyon chelicuti*, of Africa, uses ready-made holes and may even dispossess swallows of their nests under eaves.

Kingfishers dig a hole by repeatedly flying at one spot on the bank, loosening some soil with their bills each time. When they have formed a ledge, they can perch and dig more rapidly until the tunnel is 1½–3 feet (45–90 cm) long. The six or seven spherical white eggs are laid on the floor of the tunnel and incubated for about 3 weeks. During this time a stinking pile of fish bones and droppings piles up around the clutch, a squalid contrast to the magnificent plumage of the adults.

Until Ron and Rose Eastman made their prizewinning film *The Private Life of a Kingfisher* in 1966, it was believed that pieces of dismembered fish were fed to the young. The Eastman's patience and technique, however, showed the young inside the nest burrow swallowing whole fish almost as big as themselves, the bones being later regurgitated. The chicks, which are confined to the tunnel for 3–4 weeks, hatch naked. They soon acquire a covering of bristlelike wax sheaths, which are shed to reveal the adult plumage just before they leave the nest.

Halcyon days

The kingfisher has been the subject of many legends, some romantic and some prosaic. In the 12th century it was thought that not only did the bird not decay when dead but also that the corpse, if strung up by the bill, molted each year to reveal a fresh plumage. The odor of this miraculous corpse was said to be pleasant and to ward off moths. It was claimed, too, that a dead kingfisher suspended from a string pointed toward north like a compass needle or, according to another version, toward the way of the wind. The habits of kingfishers, according to the ancient Greeks, were most remarkable. Their Greek name was *halkyon*, literally meaning "conceiving at sea." It was thought that the female fed and conceived at sea and laid her eggs at midwinter in a floating nest that was so hard that it could not be cut by iron. She was supposed to have incubated the eggs for 2 weeks and fed the chicks for another 2 weeks. Being in the favor of the gods, the weather was kept calm for this period at midwinter. From this charming myth comes the expression "halcyon days," describing a period of peace and calm.

A common kingfisher flies up into the air after an unsuccessful dive for fish. This species hovers just above the water before diving in, and uses its wings to "fly" through the water after prey.

KING PENGUIN

THE KING PENGUIN LOOKS much like the emperor penguin, *Aptenodytes forsteri*, to which it is closely related. It has the same stately walk, with its long, bladelike bill held up. The king penguin is the smaller of the two, 3 feet (90 cm) long instead of almost 4 feet (1.2 m). Both species have bluish black upperparts and a white belly with yellow and orange patches around the neck, although in the king penguin the patches are isolated as two comma shapes on the side of the neck with a bib of yellow on the breast.

The king penguin lives farther north than does the emperor penguin, in the ice-free subantarctic seas from the Falkland Islands southward to the South Sandwich Islands and Heard Island. There are small colonies on Staten Island, near Cape Horn, and on the Falkland Islands off Argentina. The largest colonies are to be found on islands such as South Georgia, Kerguelen, Macquarie and Marion.

A pair of king penguins displaying to each other in one of the species' large breeding colonies on South Georgia Island.

Feeding at sea

Like other penguins, king penguins live at sea when they are not breeding. They sometimes swim long distances, traveling more than 550 miles (900 km) to feeding grounds, and even visit the fringes of the Antarctic pack ice. The latitudes in which the king penguins live are those of the roaring westerly gales, but these are unlikely to affect the penguins much except to drive them a little off course.

Penguins are superbly adapted to life at sea. Their bodies are streamlined, and a layer of fatty blubber under the skin insulates them from the cold water. The king and emperor penguins can dive to considerable depths to hunt squid and fish, which they catch in their sharp bills. The eyes of aquatic animals, penguins included, are adapted to see underwater. Light is not bent as much when it passes from water into the eye as when it passes from air. To compensate, the lens is very strong. In consequence, aquatic animals are often shortsighted on land.

Prolonged nursery period

The king penguin has the same problem of "child care" as the emperor penguin. Both are large birds and their chicks take a long time to grow, yet the Antarctic summer is very short. The emperor penguin solves the problem by starting the 7-month nursery period in midwinter, so that the chicks reach independence before the following winter. The king penguin does things differently. It lives farther north, where the sea does not freeze over and the adults are able to feed near the colony. Instead of laying its egg in midwinter, the female king penguin lays in spring or summer (between November and March), and when the chick hatches after 7½ weeks it is fed throughout the following winter, becoming independent the next summer.

Just before they commence breeding, king penguins come ashore to molt. They spend 2 weeks ashore shedding their old feathers to reveal the brilliant new coat and then retire to sea to feed and build up reserves of food before breeding. Returning to land, they make their way to the colony among the tussock grass and mud, where each male takes up his position and advertises for a mate. He stretches his neck, ruffs out his feathers, tilts his head back and calls, braying like a donkey. If an unmated female

KING PENGUIN

CLASS **Aves**

ORDER **Sphenisciformes**

FAMILY **Spheniscidae**

GENUS AND SPECIES ***Aptenodytes patagonicus***

WEIGHT
20–33 lb. (9–15 kg)

LENGTH
Head to tail: 37½ in. (95 cm)

DISTINCTIVE FEATURES
Silvery gray upperparts with blackish border from throat along flanks; mainly white underparts; yellowish upper breast; black brown head with orange patches on sides

DIET
Mainly fish; also squid

BREEDING
Age at first breeding: 5–7 years; breeding season: eggs laid November–March; number of eggs: 1; incubation period: 50–55 days; fledging period: 10–13 months; breeding interval: more than 1 year

LIFE SPAN
Not known

HABITAT
Cold seas and oceans; nests on fairly flat beaches free of snow and ice

DISTRIBUTION
Breeds on coasts of subantarctic islands at latitudes no higher than 60° S; feeds in offshore waters

STATUS
Locally common after years of persecution; for example, 240,000 to 280,000 pairs nest on Kerguelen Islands; largest colony, numbering 300,000 pairs, is on Île aux Cochons in Crozet Islands

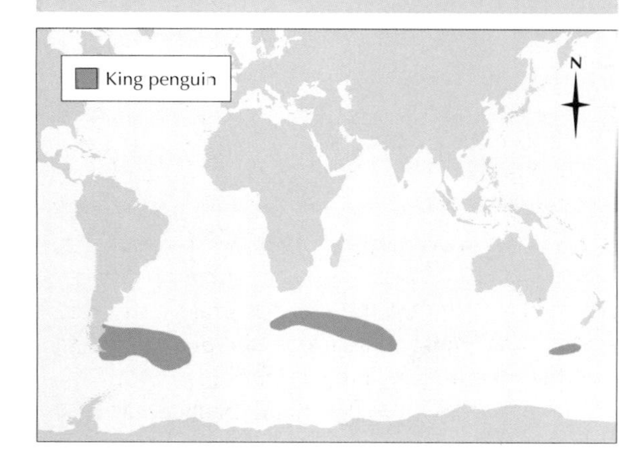

Several pounds of semidigested food are transferred at each feed, and the chicks put on weight rapidly. In the foreground of the above photograph a snowy sheathbill, Chionis alba*, is waiting to eat any scraps.*

hears him, she wanders over and the two penguins introduce themselves by flagging their bills up and down. They set off on an "advertisement walk," strutting along on their toes, waving their heads from side to side, showing off their bold patches of color. The colors are an important factor in selecting and acquiring a mate.

At first these partnerships do not last long. The male displays at any female and keeps company with a succession of prospective mates. Gradually, though, he pays attention to one in particular, the bond between them strengthens and they perform another display. Standing side by side, they raise their bills and stand on their toes as if stretching themselves.

The first eggs in the colony are laid in November, and more are laid until March. After laying a single large egg, the female goes off to feed and make up the food reserves she lost while forming the egg. The king penguin makes no nest. Instead, the male balances the egg on his feet, where it is protected by a fold of skin. He does, however, defend a small territory, which extends only as far as he can reach while incubating. The male is left guarding the egg until the female returns some 2 weeks later. Thereafter, there is a shuttle service, each parent taking a turn in guarding the egg or chick.

As the chicks in the colony grow, they spend more time on their own. Eventually they form crèches in which they huddle together while the parents go fishing. On its return a parent king penguin finds its chick by sound. It walks up to the crèche and calls, and one chick out of hundreds replies. They both walk toward each other, calling, and may even walk past each other, until another call brings them back. Several pounds of food are transferred at each feed and the chicks put on weight rapidly. As winter sets in, however, feeding becomes less frequent, and the chicks huddle in their crèches, protected by their thick, woolly down but gradually losing weight. Then, in spring, when food is once again abundant, the chicks put on weight, lose their down and the adult plumage emerges.

The king penguin chicks receive parental care for 10–13 months, after which they learn to fish for themselves. This is well-timed, coinciding with an abundance of prey from November to April. The young stay at sea for most of their early life, spending more time ashore as they get older and beginning to practice their courtship displays. At 6 years old they come ashore and start courting in earnest.

The main predators of king penguins are leopard seals, *Hydrurga leptonyx*, which lie in wait off the colonies. The seals find them difficult to catch, though, as the penguins have an alarm system. When a king penguin sees a leopard seal, it panics and rushes for the shore. Its flippers beat on the water's surface, and the clattering sound they make alerts other king penguins, sending them all rushing to the shore. The alarm not only alerts the penguins, it also confuses the leopard seals, which may then be able to catch only weak or unwary penguins.

Populations grow slowly

At one time humans were a far greater enemy. As elephant and fur seals became scarce through hunting, sealers killed king penguins for their blubber, which was used for tanning leather. Their eggs were taken and their skins were sometimes used for fancy clothing. A few colonies were wiped out, and others are only just beginning to recover their former numbers.

It took only a few years for the sealers to reduce the numbers in a king penguin colony to such an extent that it was not worth their while to exploit them further. The reason for this is the extremely slow rate of breeding. After the single egg has been laid, a pair of king penguins spends a year incubating, guarding or collecting food. Successful early layers of one season may become late layers in the next. By the time the late layers are free of their offspring, it is too late to begin again, and so they leave the colony to feed during the winter, breeding again in the spring. For this reason, a pair may lay only two eggs in a three-year period.

Consequently, king penguins, like the larger albatrosses, which also spend their first winter on the nest, cannot raise more than one young every two years. Furthermore, not all their offspring survive the first winter. If the egg is laid too late in the summer, the chick will not have had time to accumulate enough fat on which to survive the winter.

King penguins shuttle between their feeding grounds and breeding colonies in groups. Sticking together like this provides some protection against leopard seals.

KING SNAKE

The king snakes of North America are harmless to humans, as are most members of their large family, the Colubridae. A special feature of king snakes, and the origin of their common name, is their ability to prey on other snakes, even the venomous rattlesnakes. Another feature is that they are very variable in color and pattern. King snakes occasionally grow to 6 feet (1.8 m) in length, but most specimens are much smaller than this.

In many parts of their extensive range king snakes are also known as chain snakes, because the white or yellow markings on the sides of the body are looped rather like the links of a chain. The subspecies *Lampropeltis getulus getulus,* which is found from New Jersey to northern Florida and westward to the Appalachian Mountains, frequently has this pattern. The scales are smooth, and so the black background color has a shiny appearance. The underside is black with white or yellow blotches. In the Florida king snake, subspecies *floridana,* each dorsal scale is yellow or cream-colored at its base and brown at its apex. In specimens from the Mississippi Basin each dorsal scale is black, but has a white or yellow spot near its center; these are sometimes called salt-and-pepper snakes. King snakes from the western part of the United States, including the subspecies *californiae* from California, Nevada and Arizona, tend to have pronounced white or yellow bands, each of which broadens as it extends downward toward the flanks. The desert king snake, subspecies *splendida,* is often brown with black markings on the back. Finally, king snakes from the central states may have very little pattern on the dorsal surface: from a distance they appear uniform black or brown (the black king snake, subspecies *niger*).

The genus *Lampropeltis* also includes the milk snakes. The common milk snake, *L. triangulum,* which is found in most central and eastern parts of the United States, is basically black but has bands of white or yellow alternating with large, almost circular blotches of red on the flanks. Confusion is caused by the fact that milk snakes from western states are usually called king snakes: the California king snake, *L. zonata,* and the Arizona king snake, *L. pyromelana,* resemble common milk snakes more closely than do any of the varieties of king snake described above.

The Louisiana milk snake, *L. triangulum amaura,* is a mimic of the venomous coral snakes, genus *Micrurus* and *Micruroides,* and is sometimes called the false coral snake.

Terrorizing the rattlers

A king snake, which is especially active during the afternoon and evening, does not pursue other snakes. It preys mainly on small mammals, usually rodents, as well as on lizards and frogs,

***The color and pattern of king snakes vary widely between species and from region to region. Pictured is the Arizona mountain king snake,* Lampropeltis pyromelana.**

The kinkajou rarely leaves the treetops of its forest home. It is mainly active at night, sleeping by day in holes in tree trunks.

flesh and dropping the nut untouched to the forest floor, where it may later germinate. The benefit to the kinkajou is a gradual proliferation of its favorite fruit trees.

Blind young

A number of kinkajou pairs have bred in zoos in Europe and America, but little is known of their breeding except that young are often born in summer after a 3-month gestation. Usually there is a single young, although twins have occasionally been recorded. The babies are born blind, with a soft coat of black fur. Their eyes open at 10 days, and they can hang by their tail at 7 weeks. The baby normally clings tight to the fur of its mother's belly. During its early days the mother leaves it in a tree hollow when she goes foraging, but carries it from one nest to the next when necessary.

Male kinkajous are themselves ready to breed at 18 months of age, whereas females seldom breed before 2 years and 3 months. Kinkajous are fairly long-lived for a mammal of their size, at least in zoos, the maximum age so far reported being just over 30 years.

Popular pet

Its acrobatics and appealing appearance make the kinkajou a popular, if arguably unsuitable, pet. Said to be mild and docile and with the potential to be affectionate, it can, if handled carelessly, snarl and bite. Although the species is still common in the wild, pressure on its habitat is threatening localized populations.

KINKAJOU

CLASS **Mammalia**

ORDER **Carnivora**

FAMILY **Procyonidae**

GENUS AND SPECIES ***Potos flavus***

ALTERNATIVE NAMES
Honey bear; nightwalker; potto (not to be confused with the potto, *Perodicticus potto*, of tropical African forests)

WEIGHT
3–10 lb. (1.4–4.5 kg)

LENGTH
Head and body: 15¾–30 in. (40–75 cm); shoulder height: up to 10 in. (25 cm); tail: 15½–22½ in. (39–57 cm)

DISTINCTIVE FEATURES
Small, round face; small, rounded ears; large eyes; elongated body; short legs; long, prehensile tail; fluffy, honey-colored coat

DIET
Mainly fruits, flowers, eggs and honey; also insects and small vertebrates

BREEDING
Age at first breeding: 18 months (male), 27 months (female); breeding season: probably all year; gestation period: 112–120 days; number of young: usually 1; breeding interval: not known

LIFE SPAN
Up to 23 years

HABITAT
Tropical forest, especially rain forest

DISTRIBUTION
Southern Mexico south to Brazil

STATUS
Reasonably common in much of range

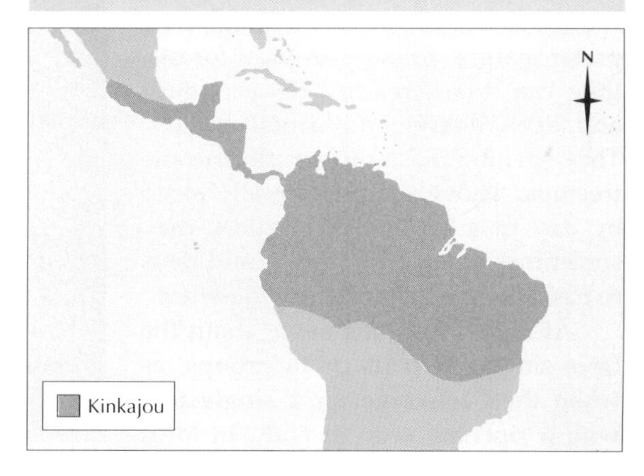

KISSING GOURAMI

This popular aquarium fish has achieved popularity for a behavior that looks uncannily like a familiar human act, but for which it would have remained in relative obscurity. "Kissing" is by no means confined to this single species of gourami, which has been chosen to show an interesting facet of animal behavior.

There are several species of gouramis, growing to 12 inches (30 cm) or more in length. All are from Southeast Asia, where they are used for food. The kissing gourami may grow to a similar size, but when kept in an aquarium it usually remains much shorter. Its body is flattened from side to side, oval in outline, with a pointed head ending in a pair of thickened lips. The dorsal and anal fins are long and prominent, both sloping upward from front to rear. Normal body color ranges from silvery green to yellow, but it is countershaded from dark uppers to paler underparts. There are dark stripes on the flanks, and a fish in good health sports a pattern of dark, crescent-shaped bands on the fins.

Thick lips for breathing and eating

Though the kissing gourami can breathe underwater through its gills, like most other fish, its gills also possess an accessory breathing organ for taking in air at the surface, permitting it to rise occasionally to gulp air. This means that it can survive in water that is slightly fouled. The kissing gourami also feeds at the surface; and the thickened lips, protruding some way from the mouth, give an advantage in these two respects.

The diet comprises animal and plant matter. In the aquarium kissing gouramis eat dried shrimps and powdered oatmeal, water fleas and dried spinach. To some extent they feed on the small algae that grow on the tank glass.

Mystery of the kiss

It is not clear whether the celebrated "kiss" is an aggressive tactic or part of the courtship behavior. The truth is probably a combination of the two. When several kissing gouramis are kept together in one aquarium, the larger of them pester the smaller by sucking at their flanks. They do the same to fish of other species. This is probably aggressive behavior. When a pair are together, however, they can be seen to face one another, swaying backward and forward, as if hung on invisible threads, and then they come together, mouth to mouth, their thick lips firmly placed together in an exaggerated kissing action. As a prelude to mating the two fish swim

Kissing gouramis are famous for a kissing gesture that looks uncannily like the familiar human act. It is part aggression and part courtship.

Kissing plays an important part in the kissing gourami's courtship. As a prelude to mating, male and female circle one another before coming together to kiss.

around and around one another in a circling movement, after which they again come together to kiss. The male then entwines himself around the body of the female during the mating act.

Left to their own devices

Little is known about the breeding habits of kissing gouramis. A mated pair builds no nest in the water; instead, the female lays 400 to 2,000 floating eggs. The adults appear to ignore these and, later, they also ignore the young, which hatch in about 24 hours. The baby fish eat ciliated protozoans for their first week, taking water fleas after that and graduating to the mixed diet as they grow older. They themselves begin to breed when 3–5 inches (7.5–12.5 cm) long.

Paying lip service

The use of the mouth as a test of strength in fighting is frequently seen in aquarium fish, especially among cichlids and gouramis. It is common, especially among the smaller freshwater species, for one fish to butt another with its mouth during courtship, and it seems likely that the mouth-wrestling and butting lead to the kissing. In his book *Tropical Aquarium Fish,* A. van der Nieuwenhuizen takes the view that in the platinum acara, *Aequidens latifrons,* a freshwater fish of Central and South America, mouth-wrestling is used to defeat a rival as well as to court a mate. He maintains that when a pair indulge in a bout of mouth-wrestling that ends in a stalemate, this means the two are physically and psychologically suited and the chances of their breeding are high.

KISSING GOURAMI

CLASS **Osteichthyes**

ORDER **Perciformes**

FAMILY **Helostomatidae**

GENUS AND SPECIES ***Helostoma temminckii***

ALTERNATIVE NAMES
Green kisser; pink kisser

LENGTH
Up to 12 in. (30 cm)s

DISTINCTIVE FEATURES
Deep, oval body; slightly pointed head with thick, protuberant lips; yellow to greenish silver overall, with paler underbelly and upper body darkening to olive green; dark horizontal stripes on flanks; greenish or dull yellow fins; usually a pattern of dark crescent-shaped bands on the fins

DIET
Mainly algae; also a range of plants and small animals, including zooplankton

BREEDING
Age at first breeding: usually when 3–5 in. (7.5–12.5 cm) length; lightweight eggs and, later, fry (young) float on water's surface

LIFE SPAN
Up to 15 years

HABITAT
Warm freshwater systems with sluggish water flow and thick vegetation

DISTRIBUTION
Tropical Asia: Thailand, Malaysia and Indonesia

STATUS
Common in native range; large quantities caught for export as aquarium fish to Japan, Europe and North America

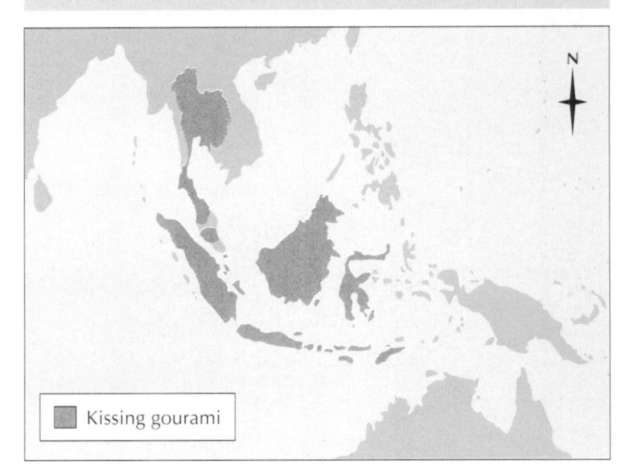

KITE

THE KITES ARE BIRDS OF PREY that have a buoyant, smooth gliding flight, after which artificial flying devices are named. Kites belong to the family Accipitridae, which also includes the hawks, harriers, eagles and vultures. Among their closest relatives are several fish eagles of the genera *Haliaeetus* and *Ichthyophaga* and the Everglade snail kite, *Rostrhamus sociabilis*, all of which are discussed elsewhere. Attention is given here mainly to the true kites of the genera *Milvus* and *Haliastur*. These have long, pointed wings angled at the wrists and a forked tail.

True kites

Both sexes of the red kite, *Milvus milvus*, are 24–26 inches (60–65 cm) long. The back and wings are reddish brown with lighter borders to the feathers, the underparts are rusty red with dark streaks and the head is grayish, nearly white in old birds. In flight red kites show a conspicuous white patch under each wing, contrasting with black wingtips, and strongly forked tails. Black kites, *M. migrans*, are slightly smaller and darker overall, and their tails are less forked.

The breeding range of the red kite extends northward from Portugal and Spain to southernmost Sweden. It also includes parts of Wales, Italy and eastern Europe, northwestern Morocco, the Canary Islands and the Cape Verde Islands. The black kite is far more common and ranges over most of Europe and Asia except for the extreme north, Africa except for the Sahara Desert and much of Australia.

The brahminy kite, *Haliastur indus*, ranges from India east to China and south to the Solomon Islands and Australia. This is the sacred kite of India, 17¾–20½ inches (45–52 cm) long, living in swamps, beside rivers and lakes and along coasts. Adults are cinnamon brown with a pure white head, shoulders and breast.

Other kites

Another group of kites, sometimes called the white-tailed kites, somewhat resembles the true kites but differs in habits. Species are found in America, Africa, southern Asia and Australia. Typical of these is the black-shouldered or black-winged kite, *Elanus caeruleus*, of Africa and

In flight the red kite shows its long, angled wings, forked tail and bold white patches under each wing. The black kite has a darker plumage, with fainter wing patches.

A black kite bathing. The black kite is common throughout its huge range, and is the most abundant bird of prey in Africa and Japan.

southern Asia, from Pakistan east to the Philippines. It is very slim and elegant, with a large, rather owl-like head. Its plumage is pale gray and white with black markings on the forward edge of the wing. The white-winged kite, *E. leucurus,* is a similar species that occurs in Central and North America, as far north as California and southeastern Texas.

Masters of soaring flight

The red kite lives in wooded river valleys, but it is sometimes seen in deciduous forests, most often at their fringes. It may spend long periods perched on a branch, always alert, descending to the ground when it sees food. It is quite active on the ground, hopping rather than walking, and having found food, the kite usually returns to a perch to eat it. The red kite also hunts over heaths, scrubby hillsides and rough pasture.

In the air red kites appear more buoyant than buzzards and hawks as they drift high over the valleys. They soar, glide and occasionally hover. Black kites are almost as elegant on the wing, but occur in a wider variety of habitats, from lowland plains to marshes, open woodland, scrub, steppe, hills and city centers. Only thick forest, hot desert and high mountains are avoided.

No food refused

Red, black and brahminy kites have a justified reputation for being scavengers and will take any dead animal food lying around, including offal

BLACK KITE

CLASS **Aves**

ORDER **Falconiformes**

FAMILY **Accipitridae**

GENUS AND SPECIES ***Milvus migrans***

ALTERNATIVE NAMES
Black-eared kite; large Indian kite; pariah kite; yellow-billed kite

WEIGHT
1⅖–2 lb. (650–940 g)

LENGTH
Head to tail: 21½–24 in. (55–60 cm); wingspan: 5¼–6 ft. (1.6–1.8 m)

DISTINCTIVE FEATURES
Medium-sized bird of prey with long, pointed wings and gently forked tail; dusky brown overall; faint whitish patch under each wing

DIET
Mainly carrion, offal and scraps; also small mammals, reptiles, birds, fish and insects

BREEDING
Age at first breeding: usually 2–3 years; breeding season: eggs laid March–June (temperate Eurasia), dry season (tropical Africa), August–December (South Africa); number of eggs: usually 2 or 3; incubation period: 26–38 days; fledging period: about 42 days; breeding interval: 1 year

LIFE SPAN
Up to 22 years

HABITAT
Anywhere except forest, hot desert and high mountains; often in urban areas and wetlands

DISTRIBUTION
Most of Europe, Africa and Asia (except north, Sahara and central China); parts of Indonesia, New Guinea and Australia

STATUS
Very common

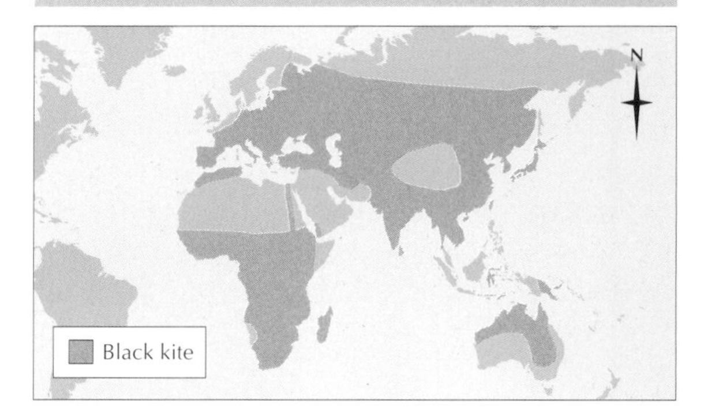

and garbage. The brahminy kite scavenges scraps from open-air markets, harbors and refuse dumps. Dead fish and fish offal are particularly favored. The black kite is bold enough to snatch fresh food from market stalls and even from baskets on people's heads.

The true kites have relatively short, weak feet for birds of prey. They are adapted to scavenging, rather than killing and carrying live prey. They will, however, take a range of small animals, from baby rabbits to mice, small birds, lizards, snakes and insects. Grasshoppers especially are picked up on the ground, and beetles are taken on the wing. In Africa the black kite is noted for preying on swarming locusts and hunting insects driven out by grass fires.

Other kites hunt in a variety of ways. The black-shouldered kite, for example, feeds mainly at dusk and dawn, quartering the ground at low levels like a harrier. It hovers frequently, dropping to the ground to catch large insects, lizards, snakes and rodents.

Garbage-filled nests

As with all large birds of prey, courtship is marked by soaring flights and mewing calls. Red kites are said to pair for life. In temperate Eurasia, each pair begins to court in earnest in about March. They sail at great heights, circling around each other or gliding low over the treetops. Sometimes one of them will fly low with a stick or wisp of wool in its talons, the other will follow and the "plaything" will be passed from one to the other, usually ending up in the nest. The nest is a platform of sticks consolidated with earth, close to the main trunk of a tall slender oak or pine. It is lined with wool, moss and an assortment of rubbish, such as hair, paper, rag, grass, dung, bones and fur. The two or three white eggs, lightly marked with brown, are laid in mid-April and incubated by only the female for about a month, the male bringing food to her. He continues to do this for a while after the eggs have hatched and then the female joins him in hunting, both bringing food to the fledglings until they are 2 months old.

Black and red compared

The black kite has a similar life history except that it also nests on the ledges of cliffs and buildings. It is more sociable than the red kite, and in parts of Africa and Asia it nests in loose groups of up to 30 pairs. Outside the breeding season it gathers in flocks at places such as refuse dumps. Most black kites and some red kites migrate, mainly from the northern parts of their ranges to more southerly latitudes with the approach of winter. The red kite, however, does not move farther south than the Mediterranean region, whereas European black kites tend to winter in tropical Africa.

Decline of the red kite

The red kite was once a common sight over city streets right across the British Isles. It featured in several Shakespeare plays. In the 18th century, however, improved hygiene—especially the introduction of drains—cut off the bird's urban food supply. Wrongly accused of preying on gamebirds and lambs, the red kite was persecuted by gamekeepers and farmers in rural areas. Tens of thousands of kites were shot, trapped and poisoned. As a scavenger, the red kite is exceptionally vulnerable to poisoning.

The red kite had vanished as a breeding bird from both England and Scotland by 1890. At one point, only four pairs were left in Wales. However, legal protection and a range of conservation measures were introduced during the 20th century. The British population of red kites reached almost 200 breeding pairs in 1998: an increase of 600 percent in just 20 years.

Brahminy kites hunt along rivers, coasts and the shores of lakes in India, Southeast Asia and Australia. The cinnamon and white adults are among the most distinctive birds of prey in the region.

KITTIWAKE

Slim and small bodied, kittiwakes are true ocean-going birds. They often wander far from the shore and in summer breed as far north as the Arctic pack ice.

The black-legged kittiwake, *Rissa tridactyla*, is a small gull that, unlike most other gulls, frequents open water and nests on cliff ledges. It is small, about 12–15¾ inches (30–40 cm) long, and resembles the mew gull, *Larus canus*, but has no white tips to the black patches on its wings. The body is white apart from pale gray on the wings and back, but in winter the gray extends up the neck to the crown and there is a sooty gray smudge behind the eye. The bill is yellow and the legs are black. Young kittiwakes have a black band across the back of the neck and a zigzag marking across the wings.

Black-legged kittiwakes live in the North Atlantic and Pacific Oceans, breeding as far south as France, Newfoundland and Alaska. To the north, they breed in Spitzbergen, Franz Josef Land and Severnaya Zemlya, all of which lie at the summer limit of Arctic pack ice. A second species, the red-legged kittiwake, *R. brevirostris*, breeds in the Bering Sea and the North Pacific as far south as the Kurile and Aleutian Islands. As their name suggests, red-legged kittiwakes have bright red legs and feet; juveniles lack the black zigzag of juvenile black-legged kittiwakes. They rarely stray far from their breeding colonies.

Oceanic seabirds

Outside the breeding season black-legged kittiwakes spend their time at sea in flocks of 1,000 or more. They live on the pack ice of the Arctic Sea, provided there is enough open water for feeding, and have been found within 130 miles (210 km) of the North Pole. Kittiwakes are rarely seen inland, and then usually only after a storm. Sometimes a flock is caught by a storm and driven inland, where many birds may die. There is a general southward movement among kittiwakes in the fall after the breeding season, and they return the following spring. There is also a movement around the oceans. Black-legged kittiwakes ringed off Britain have been recovered in Newfoundland, Labrador and Greenland.

Mainly plankton-eaters

Kittiwakes favor the plankton-rich surface of the ocean. They usually feed by plunge-diving: diving with wings half-folded but not completely submerging. On rare occasions they submerge completely and swim underwater with their wings. They catch a variety of surface-living animals such as crustaceans, squid, worms and fish. Unlike other gulls, kittiwakes are not usually attracted to carrion and offal, but they sometimes follow fishing boats and frequent canning factories and harbors to feed on scraps.

Nesting on cliffs

Toward the end of May or June, black-legged kittiwakes appear around coasts, where they nest in large colonies of up to 100,000 pairs. They often nest on the same cliffs as murres or guillemots (*Uria* spp.) and northern fulmars, *Fulmarus glacialis*. Occasionally they build on inland cliffs or on windowsills, which are artificial equivalents of rock ledges; some kittiwake colonies live on rocky islets. There may be competition for nest sites, and so kittiwakes nest on narrow ledges, to which a pair cling, facing inward, during courtship.

Most gulls build scanty nests, making little more than a depression in the ground with a lining of grass and other plants. However, kittiwakes build a solid nest of mud, grass or seaweed, forming the cuplike shape needed to keep eggs from falling off a narrow, often uneven ledge. The collection of nest material is a community event. Several dozen kittiwakes fly inland and settle on a piece of boggy ground or another suitable place where they can fill their bills with mud and grass and fly back to the cliffs, uttering their *kitt-i-waak* call. The birds place nest material on the ledge and trample it down, lining it with dry grass. To supplement their nest-building needs, kittiwakes regularly steal material from their neighbors' nests.

BLACK-LEGGED KITTIWAKE

CLASS **Aves**

ORDER **Charadriiformes**

FAMILY **Laridae**

GENUS AND SPECIES ***Rissa tridactyla***

WEIGHT
12½–17½ oz. (350–500g)

LENGTH
Head to tail: 15–15¾ in. (38–40 cm); wingspan: 37–47 in. (0.95–1.2 m)

DISTINCTIVE FEATURES
Small, slender gull with long wings and black webbed feet. Summer adult: pure white head, underparts and tail; pale gray back and wings with black wing tips; pale yellow bill. Winter adult: sooty gray smudge behind eye. Juvenile: black bill, collar and tip to tail; broad black zigzag marking across wings.

DIET
Mainly fish, mollusks and plankton; also fish scraps

BREEDING
Age at first breeding: 4–5 years; breeding season: eggs laid mid-May–June; number of eggs: usually 2; incubation period: 25–32 days; fledging period: 40–46 days; breeding interval: 1 year

LIFE SPAN
Up to 20 years

HABITAT
Open seas and oceans; nests on coastal cliffs and buildings

DISTRIBUTION
Summer: northern Atlantic and Pacific Oceans. Winter: moves south as far as Mexico, Nova Scotia and North Africa.

STATUS
Common

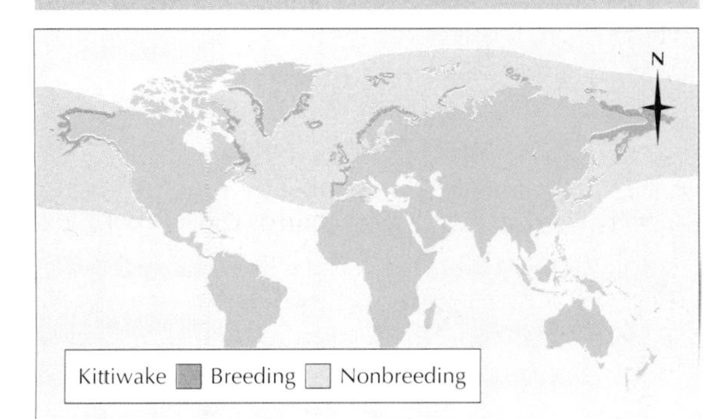

Courting kittiwake pairs bow and rub each other's bills and heads. They also face each other and utter the* kitt-i-waak *call that inspired their name.

Both parents incubate the eggs, of which there are usually two, for 25–32 days. Unlike the chicks of other gulls, kittiwake chicks cannot run about and have to spend their first few weeks in the nest. Consequently, they cannot practice flying as vigorously as other young seabirds do, or run away if attacked by their nestmates or adult kittiwakes. If they are attacked, the chicks hide their heads in submission, an act that often serves to reduce the aggression of their attackers.

Life on the ledge

Gulls that nest on open ground often fall prey to predators such as foxes, rats or ermines, but by nesting on inaccessible cliffs kittiwakes are safe from such attacks. Many of their breeding habits have altered to fit this way of life. Because of the safety of the cliffs, kittiwakes rarely give alarm calls and stay on their nests until predators are quite close. Unlike the chicks of other gulls, kittiwake chicks are not camouflaged and the parents do not remove eggshells or droppings from the nest to make it less conspicuous. They do, however, swallow or throw away waste food as a sanitary measure. At the start of the breeding season most gulls are wary of their nesting ground because of the dangers of predation, and pairing takes place away from the nesting ground. Kittiwakes, however, mate in their nests.

Some changes have been necessary for the birds to adapt to life on the ledges of cliffs. A solid nest is needed for the safety of eggs and chicks, and sharp claws and strong toes are required to maintain a grip in such a precarious position. The violent wing-beating battles typical of other gulls are impossible on a narrow ledge; kittiwakes fight by grabbing and twisting each other's bills. The chicks feed by taking food from the parents' throats. If the parents regurgitated food onto the ground for the chicks to peck, as other gulls do, it might be lost over the cliff edge.

KIWI

Kiwis, which are native to New Zealand, are the smallest flightless running birds in the Southern Hemisphere, the other runners being the emu, cassowaries, ostrich and rheas. There are three species: the common or brown kiwi (*Apteryx australis*), the great spotted or large kiwi (*A. haasti*) and the little spotted or little gray kiwi (*A. oweni*).

Each species is about the size of a domestic fowl, with a rounded body, no tail, stout but short legs, strong claws on three-toed feet and a long slender bill with slitlike nostrils at its tip. The three species vary greatly in weight, but in each the female is larger than the male. The wings are tiny, just 2 inches (5 cm) long, and are completely hidden by the hairlike body feathers that make up the plumage; the genus name, *Apteryx,* means "wingless." The eyes are small, but there are many long bristles at the base of the bill, which are probably used as organs of touch. The large ears are the chief sensory organs used in detecting intruders. The normal call of the male is thin and reedy; that of the female is more hoarse. It is a two-note call, made only at night, and sounds like *k-wee*.

Unlike all other birds, kiwis have tiny eyes and a keen sense of smell. Their nostrils are at the tips of their long bills, which is the perfect position for sniffing out prey.

Kiwis are so different anatomically from the other running birds that they are only very distantly related to them. They are more closely related to the extinct moas of New Zealand.

Forest foragers

Kiwis live in New Zealand's native forest, which once covered almost the entire country. They spend the day concealed in burrows or under the buttress roots of large trees. They are shy, retiring and hard to see in the forest because of the gloom and the birds' dark brown or gray coloring. At night they emerge to feed.

A foraging kiwi picks its way quietly and stealthily through the leaf litter, probably feeling its way to a large extent with the bill bristles. It dashes rapidly to cover at the slightest alarm, taking long strides and holding its bill out to the fore. Its main food in areas of moist ground is earthworms, plus insects and their larvae. The bill can be thrust deep in the ground, driven by the short, muscular neck, and the bill-tip nostrils enable the bird to track its prey mainly by smell. When the soil is dry in summer, the kiwi picks up fallen forest fruits and eats plenty of leaves.

Testing its sense of smell

It has always been assumed that the kiwi finds its food by smell, which is typically a weak sense among birds. In 1968 Bernice M. Wenzel of the University of California published an account in the scientific journal *Nature* of a series of experiments carried out in New Zealand. Sets of aluminum tubes were sunk into the ground in two kiwi aviaries. The tests, repeated over a period of 3 months, consisted of placing food in one tube, soil in another and a strong odorant in a third. By using different odorants and different ways of masking the contents of the various tubes, it was proved beyond doubt that a kiwi can smell food several inches down in a way that no other species of bird is able to do.

Unusually large eggs

A kiwi makes its nest in a hollow log or among the roots of a tree. Sometimes it selects a hole or burrow in a soft bank, enlarging the chamber as necessary. The hen bird then lays up to three chalky white eggs. These are very large in proportion to the hen, averaging

BROWN KIWI

CLASS **Aves**

ORDER **Apterygiformes**

FAMILY **Apterygidae**

GENUS AND SPECIES ***Apteryx australis***

ALTERNATIVE NAME
Common kiwi

WEIGHT
Male: 3⅛–6¾ lb. (1.4–3 kg); female: 4½–8½ lb. (2–3.9 g)

LENGTH
Head and body: 19¾–25½ in. (50–65 cm); female larger than male

DISTINCTIVE FEATURES
Rounded body; no tail; short, stout legs; strong claws on three-toed feet; slender, decurved bill (longer in female) with nostrils at tip; soft brown, hairlike plumage

DIET
Invertebrates, especially insects, spiders and earthworms; also fruits, seeds and leaves

BREEDING
Age at first breeding: 14 months (male), 2 years (female); breeding season: late winter to summer; number of eggs: 1 to 3; incubation period: 75–84 days; fledging period: 14–20 days; breeding interval: about 1 year

LIFE SPAN
Not known

HABITAT
Forest and shrubby areas; also farmland

DISTRIBUTION
Patchy in both North and South Islands, New Zealand

STATUS
Not threatened, despite decline in numbers in many areas

A brown kiwi foraging in the leaf litter of a New Zealand forest. The mammal-like appearance of kiwis is no coincidence: they fill an ecological niche similar to that of animals such as voles, mice and hedgehogs.

nearly 1 pound (450 g) in weight. The egg of the little spotted kiwi is 25 percent of the hen's body-weight. Incubation lasts 63–76 days in the little spotted kiwi and 75–84 days in the brown kiwi (duration for the third species is not known). In the great spotted kiwi the adults take turns to sit on the eggs, but in the other two species the male, as is more common among running birds, bears the responsibility.

The newly hatched chicks resemble small balls of soft, hairlike feathers with a spindly bill. They remain in the nest for 6 days, receiving no food during this period. Then they follow their parents on their nightly forays, finding their own food after the male has helped by clearing the ground for them. Brown and little spotted kiwis are independent at 14–20 days.

From pot to popularity

The kiwi population has decreased dramatically over the past century. The birds were prized by the native Maori as a delicacy, and their feathers were woven into cloaks for the chieftains. Then the early settlers hunted them for food. Like all New Zealand's flightless birds, kiwis suffered when dogs, cats, ermines, weasels, rats and other animals were deliberately or accidentally introduced. These alien species variously take eggs or nestlings, or both. Kiwi habitat has dwindled, too, with the inexorable spread of agriculture.

In contrast with their falling numbers, the kiwis' popularity has increased. Their image is seen on postage stamps and coins, and on the trademarks of many products from shoe polish to textiles, and their name is applied to any native New Zealander. Though the population continues to fall, at a rate of 6 percent per year, strenuous attempts are underway to preserve New Zealand's mascot and most popular bird.

KLIPSPRINGER

Bare, rocky landscapes in southern and eastern Africa are home to the klipspringer. It is very nimble-footed, leaping from rock to rock with remarkable ease.

THE KLIPSPRINGER IS ONE OF THE dwarf antelopes of the tribe Neotragini, which are all confined to Africa. It is adapted to life in bare and inaccessible rocky places. The adult klipspringer is typically just under 3 feet (90 cm) long from nose to rump with 2–5 inches (5–10 cm) of tail, and stands about 19½ inches (49 cm) at the shoulders. It weighs up to 40 pounds (18 kg). The thick, wavy coat is yellowish to reddish, with yellowish white or white on the underparts and the insides of the legs. Each hair of the upperparts is minutely banded with yellow and black, giving a salt-and-pepper effect. The hairs are bristlelike, stout and with an inner, air-filled pith unique among African antelopes.

The head is broad and triangular with a pointed snout but broad mouth, the muzzle being bare and the nostrils small. The ears are large and rounded, and their inner surfaces are noticeably ridged, suggesting efficient hearing, since ridges of this kind have been shown to direct sound waves more effectively to the eardrum. There is a prominent bare opening to the scent glands on the face in front of each eye, but there are no foot glands. Each stocky limb ends in a black-haired pastern and a cylindrical, cloven hoof 1 inch (2.5 cm) high and 2 inches (5 cm) across. The hoof is blunt-tipped with a sharp rim and almost rubbery-textured sole. The horns are short, straight spikes ringed at the base and usually worn only by males, but there is a subspecies in the Lake Tanganyika area (*schillingi*) in which the females often have horns.

The species' highly fragmented distribution across Africa has led to the evolution of up to six distinct subspecies, based mainly on differences in coat color. They range from northern Nigeria east to the Sudan, Ethiopia and Somalia, and south to the Cape.

Splendid jumpers

Klipspringers live in cliff ravines, on high rocky prominences and kopjes (rock outcrops) and in the surrounding bush, where they feed. They have been recorded on Mt. Elgon in Kenya up to 14,800 feet (13,533 m), and even on the sharp cinders at the peak of volcanic Mt. Meru in Tanzania. In East Africa's Rift Valley they shun the moister slopes of the western Rift Valley.

Though often solitary, klipspringers form small groups of up to eight, normally comprising a male with one or more females or a female accompanied by juveniles of either sex, and often kids, too. The animals are primarily active by day. The typical daily routine includes three alternating periods of rest and activity. The male habitually takes up a prominent position where he is easily visible and gives a shrill, squeaky whistle. This advertises his territory, typically 15–20 acres (6–8 ha), to his neighbors.

When disturbed, klipspringers tend to freeze, suddenly becoming near-impossible to see. When not unduly disturbed, and if only their curiosity is aroused, they give a shrill whistle. On further alarm, however, they retreat, always uphill if possible. After an initial burst of running, a klipspringer, almost always a male, pauses for a last look back before disappearing.

The klipspringer has an amazing capacity for jumping between tiny mountain ledges, hence its name, which in Afrikaans means "rock jumper." Perched on the tips of its neat hooves, a trick no other antelope practises, it is more surefooted than any goat, bouncing in zigzag leaps up a cliff like a rubber ball or descending a sheer face without losing its grip. One naturalist observed a klipspringer leap 30 feet (9 m) from a sheer cliff edge to a jutting ledge below, steady itself for a moment, then carry on running down the face.

KLIPSPRINGER

CLASS **Mammalia**

ORDER **Artiodactyla**

FAMILY **Bovidae**

GENUS AND SPECIES ***Oreotragus oreotragus***

ALTERNATIVE NAME
Cliff-springer

WEIGHT
17½–40 lb. (8–18 kg)

LENGTH
Head and body: 30½–40 in. (77–102 cm); shoulder height: 18–23½ in. (45–60 cm); tail: 2–5 in. (5–13 cm)

DISTINCTIVE FEATURES
Small size; broad head with pointed snout; prominent opening to scent gland in front of each eye; stocky legs; speckled yellow-brown or reddish coat, shading to off-white underparts; usually only male has horns

DIET
Mainly leaves and stems of bushes, shrubs and succulents; also herbs and grasses

BREEDING
Age at first breeding: 1½–2 years (male), 1 year (female); breeding season: September in some populations, otherwise year-round; number of young: 1; gestation period: 214–225 days; breeding interval: 1–2 years

LIFE SPAN
Up to about 15 years

HABITAT
Rocky mountainous habitats or open scree slopes and valleys

DISTRIBUTION
Patchy range across eastern and southern Africa, with isolated populations in Nigeria, Niger and Chad

STATUS
Conservation-dependent in most of range; endangered: Nigerian subspecies

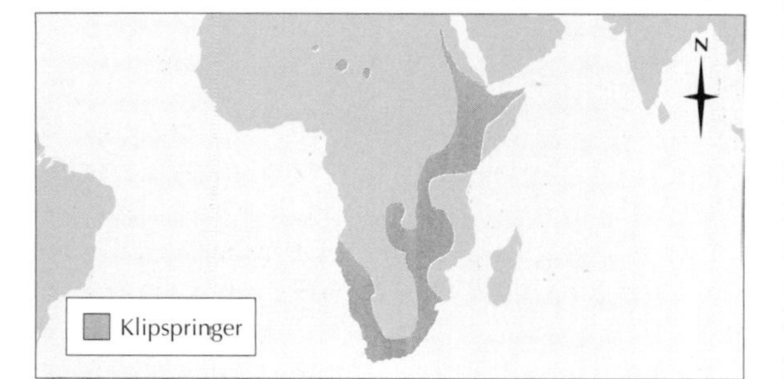

During the breeding season buck (male) klipspringers stand on pinnacles of rock with all their hoofs together. They are probably watching out for rivals.

It has been claimed that a klipspringer can leap from the ground to land on such a pinnacle no bigger at the top than a dime.

A bare living

Food for the klipspringer includes most parts of a wide variety of plants, primarily herbs, bushes, shrubs and grasses. In dry, bare habitats it survives on rock plants, especially succulents. The animal's moisture needs are low, so it tends not to visit water holes, but individuals and herds head short distances down to new flushes of greenery, where they form larger gatherings.

Statuesque bucks

The rut (mating season) is probably extended over a long period. A feature of the rut seems to be for the buck to stand watch on a pinnacle of rock with all the hoofs close together. After a gestation period of 214–225 days a single young is born, often between September and the end of January. The kid, hidden at first in undergrowth, bleats to call its mother, which visits to suckle it up to three or four times a day.

Odds against survival

A klipspringer's natural predators include leopards, large birds of prey and pythons. The only protection is the brittle and loosely rooted coat, which comes away in tufts and seems at times to confuse predators. The lost hair is soon replaced. In times past klipspringers were hunted for this hair, which was springy and light enough to be used as stuffing for saddles and mattresses. The real long-term danger to the species is its inability to adapt to ecological changes, which can lead to local extinction. The species is patchily common today, but if survival in the wild should ever become untenable, captive breeding is unlikely to prove successful.

KNIFE FISH

The term knife fish is given to certain species with a laterally flattened, tapering and bladelike body. These include the knife fish of tropical Africa, southern Asia and Southeast Asia in the family Notopteridae. Eight members of this family are also widely known as featherbacks, and are discussed elsewhere under that title. In South America there are four more families: the naked-back knife fish (Gymnotidae), the glass knife fish (Sternopygidae), the sand knife fish (Rhamphichthyidae) and the ghost knife fish (Apteronotidae).

Together the knife fish provide an excellent example of convergent evolution, in which two or more unrelated animals have come to look alike as a result of leading similar lifestyles. The Gymnotidae, Sternopygidae, Rhamphichthyidae and Apteronotidae are united in the order Gymnotiformes. Some scientists believe they are closely related to the catfish (order Siluriformes) and the characins and piranhas (Characiniformes) and, more distantly, to the carp (Cypriniformes).

Knife fish are popular among aquarists. It can, however, be difficult to tell one species from another. The various families and species are distinguished by such characteristics as the presence or absence of a pelvic girdle and electric organs, the number of scales along the lateral line and the number of fin rays and vertebrae.

Nevertheless, in all knife fish the abdominal cavity and digestive organs occupy a small part of the body behind the head, so the vent is well forward, where the pectoral fins would be in any another fish. All the fins are small, and some species, the glass knife fish and sand knife fish, have no tail fin at all. In all species only one fin is prominent: the anal fin, which runs from behind the vent along the underside of the body and is contiguous, or nearly so, to the tiny tail fin.

Life as a knife

Knife fish swim by wavelike movements of the anal fin. When the water flow is reversed, the fish moves backward with equal ease.

All knife fish live in quiet, weedy waters, in the side reaches of large rivers or in stagnant backwaters. They need to come to the surface to gulp air. In the South American knife fish the swim bladder has been transformed into a kind of lung. Studies of knife fish in Africa and Asia show that air is gulped into the gill cavity, and the spent air is later voided through the stomach, intestine and vent.

Knife fish have long bodies that are so flattened they are almost bladelike. Pictured is the black ghost knife fish, **Apteronotus albifrons.**

KNIFE FISH

CLASS **Osteichthyes**

ORDER **Gymnotiformes**

FAMILY **Glass knife fish, Sternopygidae; naked-back knife fish, Gymnotidae; sand knife fish, Ramphichthyidae; ghost knife fish, Apteronotidae**

GENUS **8 genera**

SPECIES **30 species, including banded knife fish, *Gymnotus carapo* (detailed below)**

LENGTH
Up to 2 ft. (60 cm)

DISTINCTIVE FEATURES
Eel-like body, almost cylindrical in front but drawn-out and tapering to rear; broad, oblique mouth; no dorsal, tail or ventral fins; pale grayish yellow overall, with dark transverse bands

DIET
Mainly worms, insects, shrimps and small fish; some plant matter

BREEDING
Poorly known

LIFE SPAN
Not known

HABITAT
Sluggish or stagnant fresh water

DISTRIBUTION
Guatemala south to northern Argentina

STATUS
Not threatened

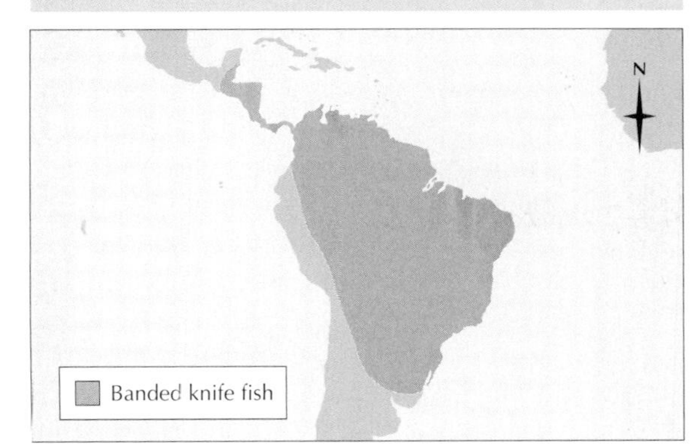

Knife fish feed at night or in twilight, on both animal and plant food. Little is known for certain about what they eat in their native habitats. In aquaria they are fed with chopped meat, worms and rolled oats, as well as smaller fish and small invertebrates such as water fleas and insect larvae. There is no obvious way of telling male from female, and little is known about their reproductive habits.

South American knife fish generate electrical pulses for locating prey and investigating their surroundings. This skill has been intensively studied in a similar fish from North and West Africa, the aba (above).

Groping in the dark

The Nile fish or aba, *Gymnarchus noliticus*, the movements and shape of which are similar to those of knife fish, has been intensively studied because of the special use it makes of electric organs. The Nile fish is discussed in greater detail in a separate article. It is of interest to note that the South American knife fish also generate electricity from organs derived from outer parts of trunk and tail muscles. Some species produce 1 to 5 Hz (hertz, or cycles per second) while at rest, rising to 20 Hz when excited. Others produce up to 1,000 Hz.

The electrical pulses set currents flowing in the water around the fish, and the pattern of the current is altered by nearby objects. Nearby animals, for instance, have a higher electrical conductivity than water, and rocks have a lower conductivity. An animal therefore concentrates the current, so increasing the current flowing through nearby parts of the knife fish's body. A rock has the reverse effect, so the fish can tell animal from mineral and food from an obstacle.

Recent tests show that in one species at least, the banded knife fish, *Gymnotus carapo*, fluctuations in the waveform of electric organ discharges enable the fish to distinguish between a friendly neighbor and a more threatening creature. The banded knife fish also responds to artificially generated water vibrations at frequencies of 125–250 Hz by raising its own electric emissions. This suggests that such fish may use their electric organs to communicate with one another.

KOALA

Koalas eat only the tender shoots of 35 species of eucalyptus or gum trees.

The koala is probably Australia's favorite animal. It is known affectionately as the Australian teddy bear, even though it is in no way related to the bears, family Ursidae. At various times it has been called bangaroo, colo, buidelbeer, koolewong, cullawine, narnagoon, native bear, karbor, koala wombat and New Holland sloth! The last two have a special interest. It was long believed the koala was most closely related to the wombat and was placed in a family of its own, the Phascolarctidae. In habits the koala recalls the slow loris and the sloth, two very different mammals that also move in a lethargic way.

Superficially resembling a small bear, the koala is about 2 feet (60 cm) high and weighs up to 33 pounds (15 kg). It has tufted ears, small eyes with a vertical slit pupil, and a prominent beaklike snout. Tailless except for a very short, rounded stump, it has thick, ash gray fur with a tinge of brown on the upperparts, yellowish white on the hindquarters and white on the underparts. It has cheek pouches for storing food, and the brood pouch of the female opens backward. All four feet are grasping. On the forefeet the first two of the five toes are opposed like "thumbs" to the rest, and the first toe on the hind foot is also opposed. Also, on the hind foot the second and third toes are fused by skin, although they do have separate claws.

Expert tree-climbers

The koala is essentially tree-living. It only occasionally descends to lick soil, apparently to aid digestion, or move to another tree. If forced to the ground, its main concern is to reach another tree, which it can do surprisingly quickly. Then it scrambles up, scaling even smooth trunks with evident ease, into the swaying topmost branches, where it clings with the powerful grip of all four feet. Although its legs are short, they are strong and there are sharp claws on the toes. When climbing a trunk, its forelegs reach out at an angle of 45° while the hind legs are directly beneath the body. It climbs in a series of jumps of 4–5 inches (10–12 cm) at a time.

Active at night, the koala spends its days sleeping curled up in a tree-fork, shunning tree hollows. The grating call of the male as he defends a territory has been likened to a handsaw cutting thin board, and is said to be one of the loudest of any Australian mammal.

Fussy feeders

The koala finds its main food in the treetops: the tender shoots of eucalyptus, 35 species of which are eaten.

KOALA

CLASS **Mammalia**

ORDER **Marsupialia**

FAMILY **Phascolarctidae**

GENUS AND SPECIES ***Phascolarctos cinereus***

ALTERNATIVE NAMES
Koala bear; variety of local names

WEIGHT
9–33 lb. (4–15 kg)

LENGTH
Head and body: 24–33½ in. (60–85 cm); shoulder height: 24 in. (60 cm)

DISTINCTIVE FEATURES
Superficially resembles a small bear; broad head with large cheek pouches; prominent beaklike snout; tufted ears; small eyes; tailless except for a very short stump; thick, woolly fur is gray above and white below

DIET
Almost entirely eucalyptus leaves, sap and bark; occasionally mistletoe leaves

BREEDING
Age at first breeding: 4 years; breeding season: September–February; gestation period: 25–35 days; number of young: usually 1; breeding interval: 1 year

LIFE SPAN
Up to 20 years

HABITAT
Eucalyptus forest

DISTRIBUTION
Mainland Australia east of the Great Dividing Range

STATUS
Near threatened; estimated population: no more than 400,000

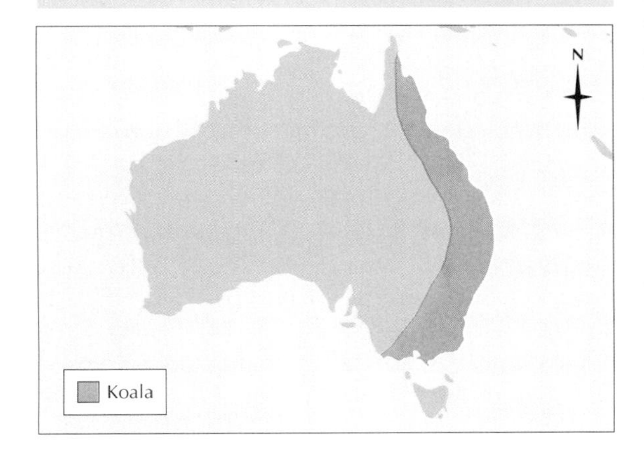

A young koala is fully furred at 6 months but stays with its mother for another 4 months after leaving the pouch.

A koala is said to smell pungently of eucalyptus. The naturalist and author Bernhard Grzimek has spoken of koalas as smelling like cough lozenges. Different races of koala rely on certain species of gum tree. Koalas on the east coast of Australia feed only on the spotted gum and the tallow wood, in Victoria only the red gum. Even then koalas cannot use all the leaves on a chosen gum. At certain times the older leaves, and sometimes those at the tips of the branches, release prussic acid, a deadly poison, when chewed. The cheek pouches, like the gut, harbor friendly bacteria that aid in breaking down the hostile diet.

However, as more and more gum trees have been felled, koalas have become increasingly hemmed in, prisoners of their specialized diet. Today special reserves have been established for the koala, but it is seldom easy to supply them with enough trees of the right kind. Koalas are

Koalas can climb even smooth tree trunks with ease. Although their legs are short, they are strong, with sharp claws on the toes.

said to eat mistletoe leaves as well, and a koala in captivity was persuaded to eat bread and milk, but without gum leaves they cannot survive.

Get off my back!

Another problem in preserving the koala is that it is a slow breeder. Usually the animal is solitary or lives in small groups. At breeding time an alpha or boss male forms a small harem, which he guards. After a gestation period of 25–35 days, each gives birth normally to one offspring. In common with other marsupials the baby is born at a very early stage of development, being just ¾ inch (1.8 cm) long and weighing no more than ⅕ ounce (5.5 g). It is immediately transferred to the pouch on its mother's belly, where it suckles and shelters for up to 7 months.

The youngster, subject of endless endearing photographs, is fully furred at 6 months but continues to stay with the mother for another 4 months after leaving the pouch, riding piggyback on her. On weaning, the young koala obtains nourishment by eating partially digested food that has passed through the mother's digestive tract. The juvenile koala is sexually mature at 4 years, and the longest-lived koala was 20 years old when it died.

Pitiless persecution

Until less than a century ago there were millions of koalas, especially in eastern Australia. Now they are numbered in the tens of thousands. In 1887–1889 and again in 1900–1903 epidemics swept through their populations, killing many. This was at a time when it was a favorite "sport" to shoot at the sleepy sitting targets. It often took several shots to finish one animal, which meanwhile cried piteously like a human baby, a fact that caused Australian naturalists to condemn the sport as the most callous. All the while, land clearance for human habitation and ranching continued to eat up the koalas' habitat.

A market also developed for koala pelts when it was found that the thick fur could withstand hard usage. In 1908 nearly 58,000 koala pelts were marketed in the city of Sydney alone. In 1920–1921 a total of 205,679 were marketed, and in 1924 more than 2 million were exported. By this time public opinion was turning against the exploitation, and before long efforts were being made to protect the surviving populations and to establish sanctuaries for them. Even today, however, koalas remain at risk from forest fires, which are a serious threat in the tinder-dry, resinous air of the hot Australian outback.

KOB

KOB ARE MEMBERS OF the genus *Kobus*, which also includes the lechwe (*K. leche*), Nile lechwe (*K. megaceros*) and waterbuck (*K. ellipsiprimnus*). Once classed as a single species, the kob is now known as two: the kob, *K. kob*, and the puku, *K. vardonii*. The kob buck (male) stands up to 39 inches (1 m) at the shoulders and usually weighs about 200 pounds (91 kg), the doe (female) being smaller and weighing about 145 pounds (66 kg). Coat color varies from orange or red to dark brown with white around the eyes and at the bases of the ears. The fronts of the forelegs are black, often with a white hoof-band, and the muzzle, lips, belly and inner thighs are white. The puku, of East Africa, is 2–3 inches (5–7.5 cm) shorter at the shoulders, the buck weighing up to 170 pounds (77 kg) and the doe up to 140 pounds (64 kg). The puku's coloring differs in details: there is no black on the legs and no white hoof-band, and the white eye-ring is narrower. Its coat is longer and rougher.

Mature male kob defend roughly circular territories from rivals. Each territory, which has good grazing and access to water, is used by the male himself and by several females.

Herds within herds

Kob follow a complicated social order. Viewed on the savanna, a large herd of kob appears to be scattered unevenly, with a few hundred here and another concentration there. Over 30 years ago, however, a ground-breaking study by Hal Buechner of the Smithsonian Institution in Washington D.C., showed that there was order in such apparent confusion. On closer inspection, the herd of kob is seen to be made up of several groups and some solitary animals, grazing or resting and chewing the cud. A few appear to roam restlessly but never transgress an invisible boundary line enclosing the territory of the herd.

In Buechner's study 15,000 kob were spread over 160 square miles (410 sq km) of a Ugandan game reserve. Within the area were 13 breeding grounds, each of which was on a ridge or knoll with good grazing, good visibility and access to water. Within it, in a central space about 200 yards (185 m) across, were 12 to 15 roughly circular territories each 20–60 yards (18–55 m) across. Some of these were touching, even overlapping, and in each was a single adult buck that spent most of his time at the center, where the grass was close-cropped and the ground was trampled. Nevertheless, males often displayed at neighboring males by walking toward the boundary with head low and feinting with their horns. Usually each male had one or more does within his territory. In addition to these circular territories, within the central space were bachelor groups of bucks and groups of unattached does.

The single individuals weaving through the herd are likely to be males without territories, eager to get one. They wander the edges of the central area, running to a particular territory and challenging the occupant or taking possession of one that is temporarily vacant because its owner has left it to go to water. The males fight with feet wide apart, heads lowered, sparring with their horns. Fights are seldom fatal; victory usually goes to the occupant male, the loser being chased well away toward the herd's outer boundary. Often he is chased or threatened by all the occupying males as he crosses one territory after another in his retreat. A buck may occupy a territory for less than a day or up to 2 months.

The love kick

Breeding is year-round, although each female is in season for only a day. On that day she leaves the unattached groups and enters the territories

Kob are dependent on regular drinking, and are tied to areas within a short walk of water.

to mate with several males. The buck tries to attract a female with a prancing display, which often carries him in his exuberance outside his own territory, perhaps into another male's territory, where he is chased off. Courtship includes what is known as the *Laufschlag* or mating kick. In this the male touches the female's underside with a stiff foreleg, placing it either under her flank from the side or between her hind legs from behind. In a great many antelopes, such as the blackbuck, gazelles, dibatag and oryxes, this is a prelude to mating and clearly indicates their relationship to each other. The antelopes that do not perform the *Laufschlag* are the tribe Alcelaphini (the hartebeest, wildebeest and impala) and the bovine group, which includes the kudu, eland and bushbuck.

Staking the claim

Since Buechner described the territoriality of the Uganda kob, similar behavior has been noted in the puku, albeit with certain differences. In one study puku territories were 8–20 times the size of those of the Uganda kob and were less rigidly observed, so when a male temporarily left his territory, his neighbor might wander in. The territory cores were not close-cropped or trampled like those of the kob, perhaps because they were so much larger, with more variety of food. The bucks' boundary displays consisted of a rapid tail-wagging, without laying back the ears. The displays sometimes ended in a chase, which might carry the bucks into another territory. Around the territorial breeding grounds were the same bachelor bands as in the kob, and here too nonterritorial males wandered in and were chased off by the owner of the territory.

KOB

CLASS **Mammalia**

ORDER **Artiodactyla**

FAMILY **Bovidae**

GENUS AND SPECIES **Kob, *Kobus kob*; puku, *K. vardonii***

WEIGHT
110–265 lb. (50–120 kg)

LENGTH
Head and body: 63–72 in. (1.6–1.8 m); shoulder height: 32–39 in. (0.8–1 m); tail: 4–6 in. (10–15 cm); male generally larger than female

DISTINCTIVE FEATURES
Long-haired, glossy brown or orange brown upperparts; white eye rings, bib and belly; black stripes on forelegs (kob only); male's coat darkens with age

DIET
Grasses

BREEDING
Age at first breeding: 2–6 years (male), 1–2 years (female); breeding season: all year, with some localized peaks; gestation period: about 270 days; number of young: 1; breeding interval: 1 year

LIFE SPAN
Up to 20 years

HABITAT
Savanna, flood plains, woodland edges, reedbeds and shrubby areas; seldom far from water

DISTRIBUTION
Kob: West Africa east to Uganda and Ethiopia. Puku: Central and East Africa.

STATUS
Conservation-dependent

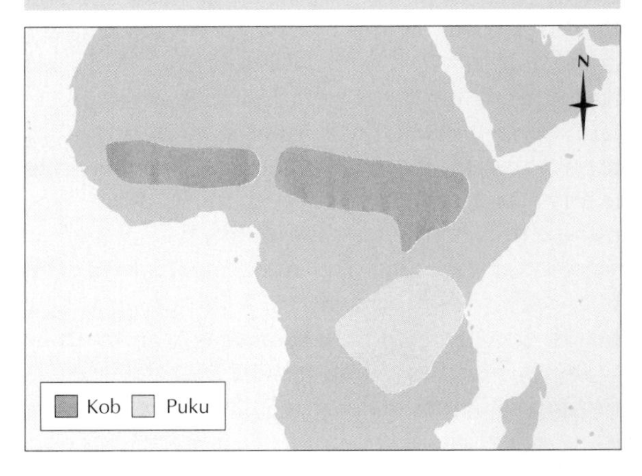

KOMODO DRAGON

Belonging to the monitor lizards, which are covered in a separate article, the Komodo dragon deserves special mention. It is not only the largest living lizard, the male growing to reach 10 feet (3 m) in length and 300 pounds (135 kg) in weight, but also the largest lizard of all time other than the now-extinct marine mosasaurs, which reached 50 feet (15 m). The only known rival to the Komodo dragon is an extinct monitor in Australia of about the same size. This lived during the Miocene period, 20–11 million years ago. Although the Komodo dragon is so large, it was unknown outside its native Indonesian home until 1912. The species is confined to a tiny area at the western end of the island of Flores, and to the three Lesser Sunda Islands: Komodo, Rintja and Padar. Komodo Island measures about 20 by 12 miles (32 by 19 km) in size. All these islands are now protected, constituting the Komodo National Park.

The Komodo dragon has a stout, somewhat flattened body, long thick neck and longish head. Its legs are short and stout and the toes have long claws. Its tail is powerful and about the same length as the head and body combined. The tongue, which is constantly flicked out of the mouth, is long, narrow and deeply cleft. Young dragons are dark in appearance with red circles all over the body and vertical bands of black and yellowish green on the neck. The neck markings disappear with age, but the red circles remain on the gray-brown bodies of adults.

A mature Komodo dragon, photographed in the wet season in the Komodo National Park, which was set up to protect the species.

Feats of gluttony

The islands where the Komodo dragon lives are hilly, their riverbeds filled only in the rainy season. The hills are covered in places with rain forest and the lowlands with tall grasses. The lizards spend the night in holes among rocks, between the buttress roots of trees or in caves. They come out at about 8:30 A.M. to look for food, chiefly carrion, which is located by smell. The tongue seems also to be used as a taste–smell organ, as in other lizards and snakes. The larger lizards monopolize any food, keeping their juniors away by intimidating them or beating them off with sideways sweeps of the powerful tail. Only when the larger individuals are full are the smaller ones able to feed.

The lizards probably kill deer and pigs as well as monkeys. They eat heavy meals that last for days. An 8-foot (2.5-m) Komodo dragon was seen to eat most of a deer, after which it rested for a week to digest the meal. Young Komodo dragons eat insects, lizards, rodents and ground-nesting birds and their eggs. Their back teeth, which are finely serrated like small saws, help them cut up flesh, although large individuals tear the meat apart with claws and teeth and swallow lumps whole. One was seen to gulp the complete hindquarters of a deer, another to swallow a whole monkey.

Middle-age spread

Mating takes place in July, and the female lays her eggs about a month later. The oval eggs, 4 inches (10 cm) long with a parchment shell, hatch the following April. Zoo specimens have grown by about 8 inches (20 cm) a year, probably reaching sexual maturity at 5 years. Up to a length of 7 feet (2.13 m) a Komodo dragon stays slender in the body. From that size, growth in length slows down markedly but there is a rapid increase in girth. Earliest reports told of dragons 23 feet (7 m) long. Although there have been more sober reports since of 12–13 feet (3.7–4 m), even the 10 feet usually quoted may be a little longer than the actual maximum measured. Reports vary greatly, and so the figure of 10 feet (3 m) is generally preferred.

A Komodo dragon feeding on a deer carcass. A heavy meal such as this will last it for days.

KOMODO DRAGON

CLASS **Reptilia**

ORDER **Squamata**

SUBORDER **Sauria**

FAMILY **Varanidae**

GENUS AND SPECIES ***Varanus komodoensis***

ALTERNATIVE NAME
Komodo monitor

WEIGHT
Up to 300 lb. (136 kg)

LENGTH
Length has always been hotly debated, but few exceed 10 ft. (3 m) and most are much smaller

DISTINCTIVE FEATURES
Huge lizard with stout, somewhat flattened body; longish head; powerful jaws with many long teeth; long, thick neck; short, stout legs with strong claws; powerful tail

DIET
Mainly pigs, deer and monkeys; some other vertebrates; most individuals scavenge, rather than hunt live prey

BREEDING
Age at first breeding: probably 5 years; breeding season: July; number of eggs: varies according to size of female

LIFE SPAN
Not known

HABITAT
Dense grassland and savanna

DISTRIBUTION
Indonesia: Lesser Sunda Islands (Komodo, Rintja and Padar) and part of Flores Island

STATUS
Vulnerable

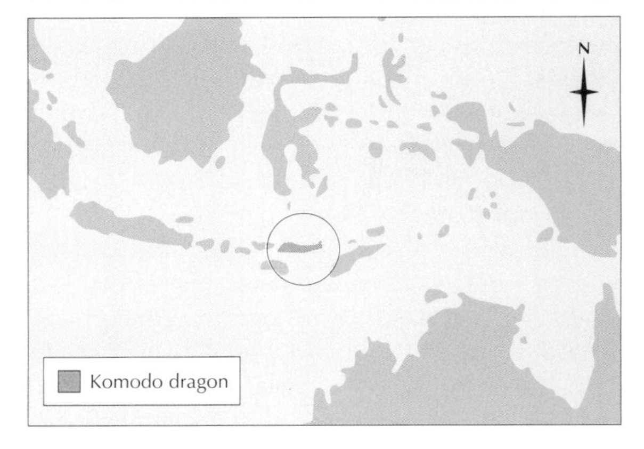

"Land crocodiles"

Komodo was an uninhabited island visited occasionally by pearl fishers and people hunting turtles. Then the sultan of the neighboring island of Sumbawa used it to deport criminals and other "undesirables." Reports began to circulate early in the 19th century of a *boeaya-darat* or land crocodile, 23 feet (7 m) long and alarmingly ferocious. In 1910 the reports became so insistent that P. A. Ouwens, director of the Botanical Gardens at Buitenzorg in Java, asked the governor of Flores to look into the reports, with the result that in 1912 Ouwens was able to publish a scientific description of this giant lizard. Then World War I broke out and the giant was forgotten in Europe, but in 1923 Adolf Friedrich von Mecklenburg, a keen explorer, went to the island of Komodo and came back with four skins of this lizard.

There are several reasons why the lizards were ignored for so long. One was that the islands were uninhabited until criminals were sent there. The stories they told were colored by their own fears and superstitions and were so exaggerated that they were disbelieved. The other reason was that it was called a crocodile, and nobody in those days, before the fashion for crocodile leather, was prepared to go all that way in search of crocodiles.

Today the Komodo dragons are a tourist attraction, the mainstay of the islands' economy. Sated on carrion, they are no longer the threat to humans they were once thought to be, although there are occasional attacks.

KOOKABURRA

Although it is a member of the kingfisher family, the kookaburra, *Dacelo gigas*, appears rather drab when compared with its brilliantly colored relatives. The famous naturalist John Gould, writing in 1844, referred to it as the great brown kingfisher.

The kookaburra is 15½–16½ inches (39–42 cm) long, its plumage a mixture of white, buff, brown and black. It is stockily built with the usual heavy head of its family, and the bill is large and heavy. The kingfisher family is divided into two subfamilies, the river kingfishers and the forest kingfishers, which usually live far from water. The kookaburra is the largest of the forest kingfishers, the female weighing up to 1 pound (nearly 500 g). Its range is eastern and southern Australia from Cape York in the north to Eyre Peninsula and Kangaroo Island. It was introduced into Western Australia in 1898 and is established in the southwestern corner. It was introduced into Tasmania in 1905.

The blue-winged kookaburra, *Dacelo leachii*, is less well known. Its range is the northern parts of Australia, north of a line from Shark's Bay in the west to southeastern Queensland. It is also found in New Guinea. It is not quite as large as the kookaburra, and is less vociferous. It is mainly distinguished by the blue in its wings.

Familiar in parks and gardens

The original habitat of the kookaburra is open forested country, where it can be seen in pairs, singly or in small groups. It has taken readily to parks and gardens and becomes friendly with people, accepting a variety of food from them. Because it is a nest-robber, it is occasionally mobbed by smaller birds, which fly at it and strike out with wings, feet and bill.

Snake-killer

The kookaburra preys on almost any animal, from large insects, crabs and other arthropods to fish, reptiles and birds. It not only robs nests of young birds and eggs but is also said to take chicks and ducklings from farms. It is a snake-killer, tackling snakes up to 2½ feet (75 cm) long, seizing them behind the head, battering them senseless or killing them by dropping them from a height. Several kookaburras may combine forces to kill a large snake.

Calling the watch

Kookaburras pair for life. The birds are strictly territorial, and a pair stays within the territory all year round. Commonly the pair is accompanied by its grown young, up to four or five in number. Each bird requires 2½ acres (about 1 ha) of land, so a group territory becomes larger with each additional young member that is allowed to remain within it. The group defends the territory together, and the young become helpers in the breeding season, incubating eggs and feeding chicks. A bird can be a helper for up to 4 years before leaving the group and setting up another territory with a mate or replacing one of its parents that has died.

The kookaburra's outstanding feature, its "laugh," comes to the fore when a family group defends its territory. Two or three birds may simultaneously deliver the call, which rises from a chuckle to an ear-splitting cacophony before once again subsiding. Gould wrote: "It rises with

The kookaburra is a specialized snake-killer. Its technique is to seize a snake behind the head and batter it senseless, then eat it at leisure.

Kookaburras are often seen in pairs or small groups. They are renowned for their chuckling "laugh."

the dawn when the woods re-echo with its gurgling laugh; at sunset it is again heard." Various naturalist authors have since referred to the regularity with which the kookaburra gives out its "shouting, whooping and laughing chorus," mainly at dawn and dusk. Shorter calls may be used in courtship, or to signal aggression or warn of danger.

The breeding season runs from September to December (spring–summer in the Southern Hemisphere). The nest is built in a hollow tree or in a hole in a bank, sometimes in a chamber tunneled out of a termites' nest. The eggs are white and somewhat rounded, and there are two or occasionally three in a clutch. When five are found in a nest, one or more is likely to have been laid by a helper.

A detailed account has been given of a nesting pair in a zoo. The nest was in the hollow base of a tree, and the birds tunneled out the soil below so that the nest was 3 inches (7.5 cm) below ground level. One egg was laid and then a second the following day and a third three days later. Male and female shared the incubating for 25 days. Several times the hen rapped on the tree with her bill. The male responded to this by going in and relieving her at the nest.

Kookaburras are vigorous in defending their nest and young. Such species that practice cooperative breeding tend to be long-lived: they are known to have survived 12 years in the wild. However, away from the homesteads and suburbs the kookaburra is threatened by the continued felling of trees and the advance of human settlement. They are also harassed by introduced starlings taking over nesting sites in hollow trees.

KOOKABURRA

CLASS **Aves**

ORDER **Coraciiformes**

FAMILY **Alcedinidae**

GENUS AND SPECIES ***Dacelo gigas***

ALTERNATIVE NAMES
Laughing kingfisher; laughing jackass; jackie; great brown kingfisher; alarmbird

WEIGHT
Male: 11–12 oz. (310–345 g); female: 12½–17 oz. (355–480 g)

LENGTH
Head to tail: 15½–16½ in. (39–42 cm)

DISTINCTIVE FEATURES
Sturdy build; fairly long, chunky bill; whitish head with dark brown ear coverts; white underparts; brown back; rufous tail with black bars

DIET
Mainly lizards, snakes, invertebrates, eggs and nestlings; also fish and feeder scraps

BREEDING
Breeding season: September–December; number of eggs: 2 to 5; incubation period: 24–26 days; fledging period: 33–39 days; breeding interval: 1 year

LIFE SPAN
Up to 12 years

HABITAT
Open, dry eucalyptus forest; woodland; city parks and gardens

DISTRIBUTION
Eastern, southeastern and southwestern Australia; Tasmania

STATUS
Very common; estimated population in mid-1980s: 66½ million birds

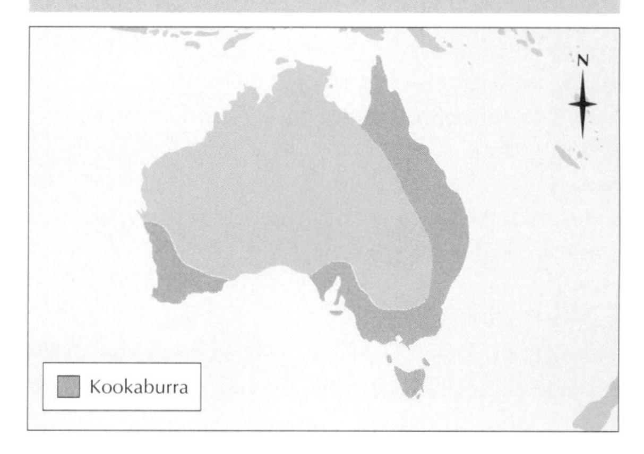

KRILL

KRILL IS A NORWEGIAN WORD for the food of the baleen whales. It was first used by Norwegian whalers in Arctic waters, but worldwide there are 85 species of crustaceans known as krill. A few of these species occur in huge swarms, and the most important, covered here, is *Euphausia superba* of the Antarctic. The interest today in krill is twofold. It has been put forward as a food source for the world's soaring population, and there is also a scientific puzzle: to discover how whales find krill.

Euphausia superba looks like a shrimp, up to 2.5 inches (6 cm) long with antennae adding another 1 inch (2.5 cm). It has a pair of stalked compound eyes and eight pairs of branched legs on the abdomen, the final pair forming the tail fan. There are also 10 light organs: one at the base of each eye-stalk, two pairs on the underside of the thorax, at the bases of the second and seventh thoracic legs, and a line of four under the abdomen. The body of the krill is translucent but has numerous blotches and spots of reddish brown. At night *E. superba* glows with light, and a shoal becomes a mass of living blue-green fire.

Restless shoals

Krill contains 7 percent fat and 16 percent protein. It is the main food of the large baleen whales in the Antarctic and of the misnamed crabeater seals, Adélie and gentoo penguins, and several other birds. It is also the principal food of squid, which in turn form the staple diet of many birds and marine mammals. Moreover, at least 32 species of fish and many other species of birds eat krill. It forms the most important part of the Antarctic zooplankton, occurring in large shoals, which, when near the surface, appear to color the water brick red. Some shoals may be a few feet across, wheras others may cover half an acre (2,000 sq m). In his *Great Waters,* Alister Hardy has described how the smaller shoals may have an indefinite outline, like that of a gorse bush, while other shoals extend in long wavy bands from one to several feet or even yards across. They may be circular, oval or oblong, often having the shape of huge amoebae. Even the smaller shoals consist of thousands of euphausians, all swimming around and around in a whirling mass that continually changes shape. At one moment they are at the surface and the sea appears blood red, at another they go deeper so the color changes to brick red, or they may go deeper still, to as much as 1,000 feet (300 m). Between the shoals are wide gaps in which no krill is to be found.

A more vivid idea of the abundance of Antarctic krill can be conveyed by a few statistics. During January to April the shoals contain 10 pounds of krill per cubic yard (6 kg per cubic meter), and before the whales were hunted they probably ate 50 million tons (45 million tonnes). A full-grown whale may eat as much as 4 tons (3.6 tonnes) of krill every day in the summer, when the crustaceans are at their most abundant, in order to build up a thick layer of blubber.

Complex life history

The spawning season begins from January to March and lasts 5 or 6 months. The sperm are oval and pass from the testis into a sac. The walls of this sac secrete a horny cuticle to enclose them within a slender-necked, flask-shaped spermatophore. At mating this is passed out, and a handlike organ on the first of the abdominal legs, the petasma, holds the spermatophore and attaches it accurately to the opening of a pouch on the female's body near where the eggs will pass out, so the sperms fertilize the eggs as they are laid. Eggs are found down to depths of about 6,000 feet (1,820 m). The hatched krill larva goes through nine changes, at each of which it increasingly resembles an adult. The first stage is known as a nauplius, which changes into a metanauplius. Then follow three calytopis stages and four

Euphausia superba, ***of Antarctic seas, is one of the most important species of krill. It is the main food of many whales, seals, penguins, squid and fish.***

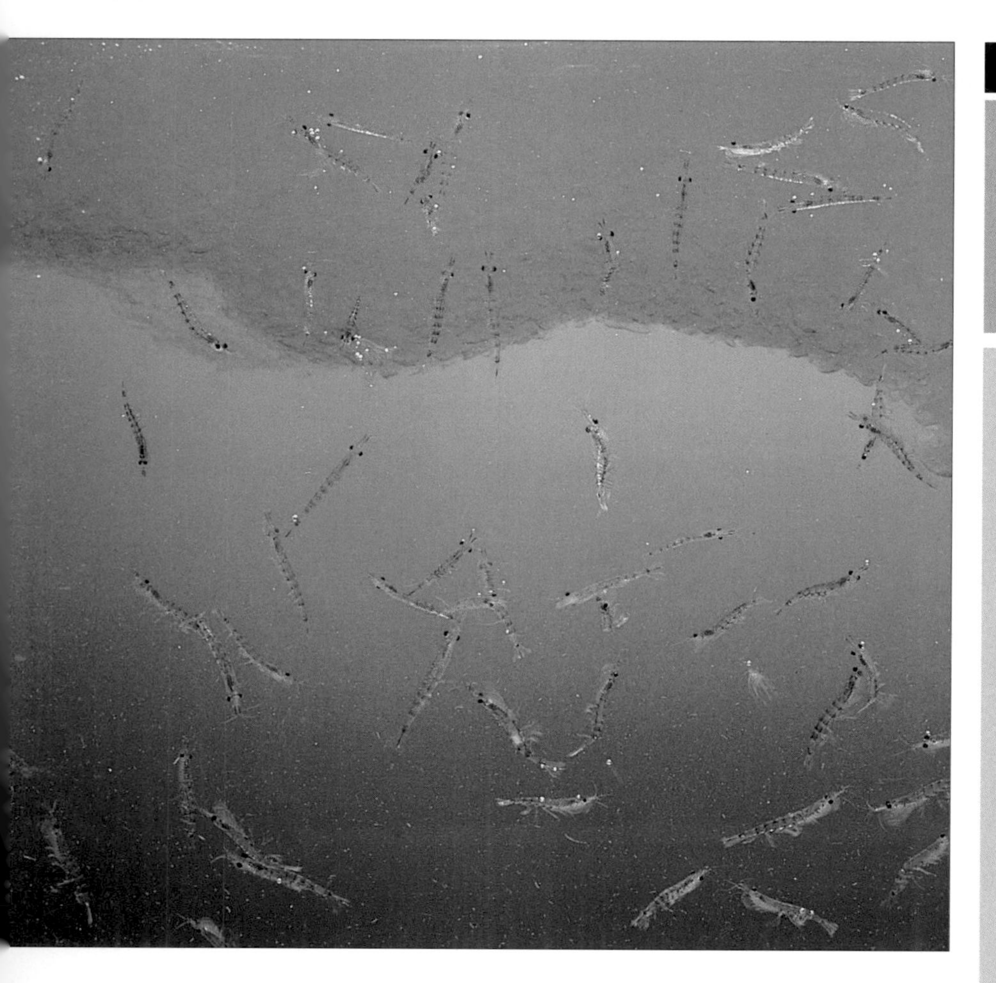

Adult krill feed on phytoplankton, which they filter from the surface waters. Both phytoplankton and krill occur in enormous quantities during the polar summer.

furcilia stages. Krill take between 2 and 4 years to reach maturity. During the course of these changes, the growing krill are moving up from the depths and reach the water's surface as the first calytopis stage.

Research on krill

Krill feed on the phytoplankton, the mass of microscopic plant organisms that flourish under the effects of solar radiation and are therefore at their busiest during the long daylight hours of the polar summer. Surprisingly, there are no krill where the plant food is densest. This may be because a bloom of phytoplankton actually poisons marine animals. Likewise, the phytoplankton are intolerant of excess ultraviolet light, the section of the waveband that is increasingly penetrating polar waters as a result of depletion of the ozone layer.

Recent work has revealed that the krill shoals are far less predictable than once was thought. Areas normally highly productive may have lean years when almost no new adults are recruited. The causes are not known in detail, though changes in the amount and extent of pack ice seem to be in some way important.

Several nations have sent trawlers to the Antarctic to try catching krill. Shoals have been located quite easily by using echo sounders. Russia and Japan are the main krill-fishing nations, but others are involved too. Rarely more than 300,000–350,000 tons (272,000–317,000 tonnes) of krill are caught in any one year. Marketing krill has always been a major problem, but there are several other concerns, including fishing rights and environmental concerns, such as the possibility of localized depletion of stocks having effects on food chains.

KRILL

PHYLUM **Arthropoda**

CLASS **Crustacea**

ORDER **Euphausiacea**

GENUS AND SPECIES ***Euphausia superba***

LENGTH
Up to 2½ in. (6 cm)

DISTINCTIVE FEATURES
Shrimplike crustacean; long antennae; 1 pair of stalked compound eyes on head; 8 pairs of branched legs on abdomen; 10 pairs of light organs; translucent body, but glows with light at night

DIET
Phytoplankton

BREEDING
Age at first breeding: 2–4 years; young pass through 9 larval stages

LIFE SPAN
Up to about 7 years

HABITAT
Open seas and oceans, mainly in top 100 yd. (90 m)

DISTRIBUTION
All around Southern Ocean

STATUS
Superabundant, but populations fluctuate

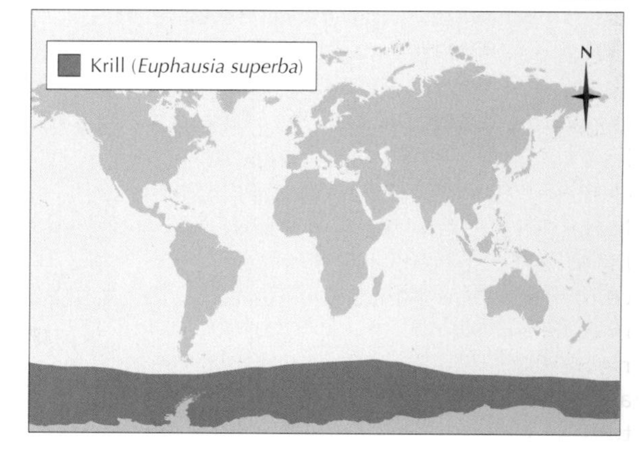

KUDU

A female greater kudu nursing her single calf. Kudu have a variety of white marks on the face and white-striped flanks, in common with other spiral-horned antelopes such as the bushbuck, nyala, bongo and sitatunga.

Kudu are among the largest of the antelopes. There are two species: the greater kudu and lesser kudu. Like their closest relatives the bushbuck and the nyala, they have spirally twisted horns, white spots and chevrons on the face and white-striped flanks. Males alone have horns, which are about 46 inches (1.2 m) long and have at least 2½ turns compared with 1½ turns in related species. The male greater kudu stands 50 inches (1.3 m) or more at the shoulders; he has a short, erect mane and a throat fringe. The coat is reddish fawn, turning blue gray in old males. There are 4 to 12 white stripes on the flanks. A male weighs up to 700 pounds (315 kg), a female up to 475 pounds (215 kg). The lesser kudu, up to 43 inches (1.1 m) high, has a scanty mane and is yellow gray with 11 to 14 white flank stripes, a white crescent on the throat, another on the chest and white spots on cheeks and nose.

The greater kudu lives in isolated hilly districts from Lake Chad, Ethiopia and Somalia, through eastern Africa to South Africa. It is more scattered than the lesser kudu, which lives in scrub and desert bush in eastern Ethiopia, Somalia and northern and eastern Kenya.

Kudu cousins

An antelope that somewhat resembles the kudu is the bongo, *Tragelaphus eurycerus*, a robust forest animal about 48–50 in. (1.22–1.27 m) high, which stands somewhere between the eland and the kudu–bushbuck group. Like the eland, it has a tufted tail, the females have horns and it lacks inguinal glands; like the kudu and bushbuck, it has horns with a fairly open spiral and it lacks the eland's dewlap and forehead "mat" of hair. The horns of the bongo have only one turn, and they have peculiar yellow tips. Its coat is reddish, with many stripes, chevrons and facial spots like the lesser kudu.

The bongo is confined to the forest belt of Africa, along the Guinea coast to Cameroon, across the Democratic Republic of the Congo (Zaire) and into some isolated highland forests, 7,000–10,000 feet (2,100–3,000 m) in altitude.

Only the male greater kudu has horns, which are spirally twisted. The horns have at least 2½ turns compared with 1½ turns in related species of antelopes.

In East Africa it lives on the Aberdares, Mt. Kenya, the Mau Escarpment and the Cherangani Hills. The bongo's habitat may be that of the ancestors of all the above species.

Group harmony

Kudu go about in small groups, usually of two to four but sometimes up to 11, and about one individual in 20 is solitary. The smaller groups may be all males or all females with their calves, the larger groups being made up of both sexes with about two females to every male. The members of a group keep a certain distance apart while feeding, but when they lie down they draw close together. They then often indulge in mutual grooming, which can be intensive, the groomers licking each other on the head and foreparts. The high-ranking animals, always males, choose which way the group is to go and when it is to lie down. Should a low-ranking animal lie down first, the dominant one will make it rise again. But when it lies down, a dominant kudu seems to lose status: younger animals and females often come and pester him, which they would not do if he were standing up.

Kudu are largely browsers, but they favor grasses during the rainy season. Their main enemies are leopards, hunting dogs and lions.

KUDU

CLASS **Mammalia**

ORDER **Artiodactyla**

FAMILY **Bovidae**

GENUS AND SPECIES **Greater kudu, *Tragelaphus strepsiceros*; lesser kudu, *T. imberbis*; bongo, *T. bongo***

ALTERNATIVE NAMES
Koodoo; grand kudu

WEIGHT
265–700 lb. (120–315 kg); male much heavier than female

LENGTH
Head and body: 6¼–8 ft. (1.9–2.5 m); shoulder height: 3¼–5 ft. (1–1.5 m)

DISTINCTIVE FEATURES
Tall build; stiff crest along spine; brown or reddish brown, with 4 to 12 white stripes across body. Male: spiral horns up to 6 ft. (1.8 m) along curve; crest extending down from throat; becomes grayer with age.

DIET
Wide range of grasses, herbs, succulents, leaves, vines, flowers and fruits

BREEDING
Age at first breeding: 5 years (male), 3 years (female); breeding season: all year, with localized peaks; number of young: 1; gestation period: about 270 days; breeding interval: 1–2 years

LIFE SPAN
Up to 23 years in captivity

HABITAT
Woodland and thickets, often near water

DISTRIBUTION
Isolated regions of sub-Saharan Africa

STATUS
Conservation-dependent

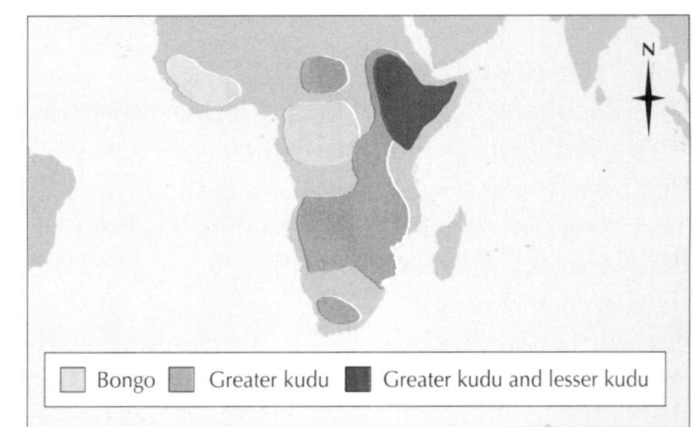

Courtship wrestling

Mating takes place at any time of the year, but there is a peak in the late part of the year south of the equator, in Zambia and South Africa, and one in the early part of the year in northern Kenya, north of the equator. Gestation is about 9 months. Little real fighting occurs in the rut, but males spar and threaten each other at any time. One male may try to catch another's head from the front, cross horns and push with his forehead. The opponents turn their heads away from each other during pauses in these wrestling matches and again at the end of the fight. The fights probably do no more than maintain or decide social dominance. Neck-fighting is also used between males, and between male and female during courtship. Males often attack trees and bushes and scrape at the ground. Such behavior that has been interpreted as a redirected threat, such as when an animal's aggressive courtship behavior is frustrated.

Although a male will not attack a female, females readily attack males or each other. Like the males they shove with the forehead, and although they have no horns, the strength with which they ram an opponent in the shoulder or flank is considerable. They may even snap an opponent's jaws, something no male would ever do. A young bull attacked by a female may threaten her but will not defend himself: an early example of chivalry!

In the courtship ritual the male and female wrestle with their necks. The male drives a female as a sheepdog rounds up a sheep: he overtakes her and brings her to a halt. He thrusts his outstretched head and neck along her back from the rear before mounting her. He displays to the female with head upstretched, while the female lifts her nose high. If she attempts to move away, the male greater kudu may plow the ground in front of her with his horns. The male lesser kudu simply runs in front of his female to stop her from escaping.

There is one calf at a birth, and it is hidden in dense cover in a place that the calf selects by instinct. Its coat is a pale cinnamon color. The mother grooms its entire body, not just the foreparts, as when adults groom each other. In its first few days she eats its feces and drinks its urine, a precaution that removes odors that might otherwise attract predators. The mother rests some distance from her calf, although always within sight of it. Whenever the calf stands up, she goes to join it.

Kudu usually move about in small groups of either males only or females with their calves. Larger groups containing adults of both sexes are less common.

LACEWING

SEVERAL INSECT FAMILIES of the order Neuroptera are given the name lacewing. The most familiar in Europe are the green lacewings, family *Chrysopidae*, most of which have a green body and metallic golden eyes. The brown lacewings, family *Hemerobiidae*, smaller and brown or gray, are quite common among bushes and low plants. The spongilla flies or sponge flies, family *Sisyridae*, are small lacewings that have aquatic larvae, which live as parasites on freshwater sponges. The largest British species is the giant lacewing, *Osmylus fulvicephalus*, which has a 2-inch (5-cm) wingspan. It is found near woodland streams. The tiny dustywing lacewings, family *Coniopterygidae*, are of economic importance as their larvae prey on the red spider mite, a serious pest of fruit trees.

Most lacewings fly at night and are attracted to artificial light. Green lacewings often fly out by day when bushes and low branches are disturbed, and the small brown Hemerobiidae species can be found by shaking leafy branches over an inverted umbrella. Some may fly up, but most will fall into the umbrella. The beautiful giant lacewing is best found by searching in culverts and under bridges where small streams run. It is a local and not very common insect. The giant lacewing and some of the green lacewings have a strong and unpleasant odor, which gives them some measure of protection against birds and other predators.

A high-speed studio photograph of a green lacewing. Lacewings fly by night, although green lacewings will flutter from cover if they are disturbed during the day.

Pincushion eggs

Lacewings undergo a complete metamorphosis, with larva, pupa and imago stages. The female green lacewing has the unusual habit of laying her eggs on the tips of long, hairlike stalks, which she makes herself in groups on leaves. She first dabs a drop of gummy liquid from the tip of her abdomen onto the leaf and then, raising her abdomen, draws it up into a slender stalk, which instantly hardens. She then lays the egg at the tip of the stalk. The larvae are predatory, feeding largely on aphids. When fully grown, each spins a cocoon of white silk given out from a spinneret, which is at the hind end of the body, not on the head as in the silk-spinning caterpillars.

The cocoons are usually attached to leaves or bark. Most of the young spend the winter as larvae inside the cocoon. They pupate the following spring, but one very common species, *Chrysopa carnea*, hibernates as an adult. Although it is green before hibernation, it turns brown soon after it settles down, becoming active and green again in the spring. These brown hibernating lacewings can often be seen inside houses in autumn and early winter.

The giant lacewing spends the whole of its larval life, and spins its cocoon, in wet moss beside streams. Those of the spongilla flies in the genus *Sisyra* are wholly aquatic, breathing through gills. Later they leave the water and spin cocoons in bark crevices and similar places. Most other lacewing larvae live among foliage.

Hypodermic feeding

In captivity brown lacewings have been seen preying on aphids. As larvae all are predators, and in gardens they

LACEWINGS

PHYLUM **Arthropoda**

CLASS **Insecta**

ORDER **Neuroptera**

FAMILY **Green lacewings, Chrysopidae; brown lacewings, Hemerobiidae; dusty-wing lacewings, Coniopterygidae; sponge flies, Sisyridae**

GENUS AND SPECIES **Many**

ALTERNATIVE NAMES
Green lacewings: golden-eyes; stinkflies. Brown lacewing larvae: aphid lions; aphid wolves. Sponge flies: spongilla flies.

LENGTH
Adult body length: ⅛–⅚ in. (3–20 mm); adult wingspan: up to 2 in. (5 cm)

DISTINCTIVE FEATURES
Small, medium or large insects, depending on species. Adult: soft body; long antennae; generally brown or green; 2 similar pairs of flimsy wings covered with delicate network of veins. Larva: flattened, louse-shaped body; sharp, calliper-like jaws.

DIET
All species: aphids and other small insects; aquatic Sisyridae larvae: freshwater sponges

BREEDING
Undergo a complete metamorphosis. Breeding season: summer (temperate regions), all year (Tropics); larval period: pass through 3 instars (stages) and usually overwinter as larvae; breeding interval: 1 year (most species), 2 or 3 generations per year (others).

LIFE SPAN
Usually up to 1 year

HABITAT
Adult: variety of habitats. Larva: some species in fresh water, others in foliage.

DISTRIBUTION
Almost worldwide

STATUS
Very common or abundant

The larvae of lacewings (adult, above) are voracious predators with caliper-like jaws. They hunt aphids and other small insects.

share with the ladybug and hoverfly larvae the feast of swarming aphids or greenflies. Lacewing larvae are flattened, louse-shaped creatures with sharp-pointed, hollow jaws resembling a pair of callipers. They mostly live among foliage and crawl actively about searching for aphids and other insects. A larva seizes its victim and pierces it with the hollow jaws, through which a digestive juice is injected in order to liquefy the body contents. The resulting soup is then sucked back by the larva. In both the injection and the suction, the jaws act like miniature hypodermic needles. The larvae of the larger species feed on small caterpillars and other insects as well as on aphids, and those of the giant lacewing eat any insects they can find. The smallest species prey on microscopic mites and their eggs, and in doing so may render as valuable a service to humans as the aphid-eaters.

Tidy diners

The larvae of some of the green lacewings set a wonderful example of what to do with the wrappings after an open-air meal. When one of them has sucked an aphid dry, it does not throw it away but holds the husk in its jaws and presses it down onto its own back, which is covered with stiff, hooked hairs. These hold the husk in place, where it dries and shrivels. After a time the larva is covered with a mass of husks, which makes it look more like a small heap of dried rubbish than a living insect.

When the larva molts its skin, the accumulation of husks is lost, but the larva starts to replace it as soon as it begins feeding again. Most of us must have seen these disguised larvae at some time or other, on the leaves of rose bushes or other foliage, but without noticing them. Almost certainly insectivorous birds also miss them when they are searching among the leaves.

LADYBUG

SMALL, BRIGHTLY COLORED beetles, oval or almost circular in outline, ladybugs were regarded with affection long before it was realized that they are useful as well as pretty. The name ladybug (sometimes ladybird or lady beetle) dates from the Middle Ages, when the beetles were associated with the Virgin Mary and called "beetles of Our Lady." Their coloration is generally red or yellow with black spots, and the pattern tends to be variable, extremely so in some species. A few, like that known as *Coccidula rufa*, are brown without conspicuous markings and are not usually recognized as ladybugs. The colorful species have a strong and bad smell, and they taste equally bad. Their bright colors doubtless serve as a warning to predators not to try to eat them. Both ladybug adults and their larvae prey on aphids, killing them in great numbers.

Easy to spot

Other than color, ladybugs are most easily identified by their spots. In Britain, for example, the four most common species are the 2-spot, 10-spot, 7-spot and 22-spot ladybugs (or ladybirds, as they are exclusively known there). The first four of these have been given the scientific names *Adalia bipunctata, A. decempunctata, Coccinella septempunctata* and *Thea vigintiduopunctata* respectively. In each the second name is a Latin translation of the number of spots on the two elytra (wing cases). Even scientists balk at long names, so these four ladybugs are usually referred to as *2-punctata, 10-punctata, 7-punctata* and *22-punctata.*

The first species is red with a single black spot on each elytron (wing case), but black specimens with four red spots are common, and the beetle is sometimes yellow with black spots. The underside and legs are black. The second species is reddish or yellow, usually with five black spots on each elytron, but the ground color may be black as in the previous species. The underside is brown and the legs yellowish. The 7-spot is larger than the first two species and its colors hardly vary at all. It is orange red with a black spot on the line dividing the elytra and three others farther back on each side. The last species is much smaller with 11 black spots on each side on a bright yellow ground.

The eyed ladybug, *Anatis ocellata*, is a large British species. It has black spots on a red ground, each spot being surrounded by a halo

A mating pair of 7-spot ladybugs. The strong coloration of adult ladybugs warns birds and other predators that the beetles smell foul and taste bad.

LADYBUGS

CLASS **Insecta**

ORDER **Coleoptera**

FAMILY **Coccinellidae**

SUBFAMILY **Epilachinae, Coccinellinae and Chilocorinae**

GENUS AND SPECIES **4,500 species worldwide, including about 400 in U.S.**

ALTERNATIVE NAMES
Ladybird (Britain only); lady beetle

LENGTH
Adult: ¹⁄₂₅–⅜ in. (1–10 mm)

DISTINCTIVE FEATURES
Adult: small or medium-sized; body usually round or oval; upperside of elytra (wing cases) brightly colored in most species and usually patterned with spots, bands or stripes; short, clubbed antennae; short legs. Larva: covered with bristles; colored black, orange, blue or red, depending on species.

DIET
Most species: aphids, or nectar and pollen when aphids are scarce; other species: plants and mildew

BREEDING
Breeding season: spring–early summer; number of eggs: usually 100 to 200; hatching period: 5–8 days; larval period: about 21 days; breeding interval: 1 year

LIFE SPAN
Usually up to 1 year

HABITAT
Almost any terrestrial habitat

DISTRIBUTION
Virtually worldwide

STATUS
Generally common

A group of ladybugs (Adalia conglomerata) hibernating on the trunk of an ash tree. Adult ladybugs emerge in summer and breed the following spring.

of yellow. The eyed ladybug is generally ⅜ inch (9 mm) in length, and it lives among the foliage of various pine trees.

Crowded winter resorts

In summer ladybugs fly actively about among foliage. In winter they hibernate as adults, often in large groups. Sometimes 50 or 100 of them can be found crowded together under a piece of loose bark, on a post or in a porch. They often congregate in houses and usually go unnoticed until they come out in spring. In California crevices and caves on certain hilltops are well known as hibernation resorts where ladybugs gather in the thousands.

Hordes and hordes of ladybugs

Ladybugs usually lay their orange-colored eggs on the undersides of leaves, in batches of 3 to 50. Several batches are laid by one female, totaling 100 to 200 eggs, sometimes more. Because the beetles themselves feed on aphids or greenflies, they tend to choose places where these are abundant in which to lay, so the larvae find food handy from the start. The eggs hatch after 5 to 8 days, turning gray shortly before they do so. The larvae are active, bristly and variously colored in patterns of black, orange, blue and red. Like the adult beetles, they feed on aphids, but since they are growing rapidly they are far more voracious. The larval stage lasts 3 weeks or so, during which time several hundreds of aphids are eaten.

When thousands of aphid-eating ladybugs are each laying hundreds of eggs, and every larva is consuming hundreds of aphids, it is evident that very large numbers of greenflies are destroyed, and the benefit to plants, both wild and cultivated, is enormous. The pupa is usually attached to a leaf. The whole life cycle takes from 4 to 7 weeks, so several generations of ladybugs may be produced in a summer, although there is usually only one generation in a year. One small group of ladybugs are not predatory but feed as larvae on plant food. A single species occurs in Britain, the 24-spot ladybug, *Subcoccinella vigintiquatuorpunctata*. It feeds mainly on plants such as campions, chickweed, trefoils, vetches and plantains. Elsewhere in Europe it is occasionally a minor pest of alfalfa.

Ladybug farms

The principle of using one species of insect to control the numbers of another is now well known and is often favored over the use of poisonous insecticides. An early example of an operation of this kind concerns the use of a ladybug. Toward the end of the 19th century the California citrus orchards were devastated by the cottony-cushion scale insect, which was accidentally introduced from Australia. A brightly colored ladybug, *Rhodalia cardinalis*, was found to be a natural enemy of the scale insect in Australia, and in 1889 some of these ladybugs were brought to California and released in the orchards. They effectively controlled the scale insect there, and they have since been introduced into South Africa. The California citrus growers were also troubled by aphids and other plant bugs, and use was made of a native ladybug, a species of *Hippodamia*, that hibernates, as mentioned earlier, in caves in the hills. The *Hippodamia* beetles were collected and sold to the citrus farmers by the liter (8,000 to 10,000 beetles in each liter) and later by the gallon. This control was started in 1910, neglected, then revived during World War II.

Even this is not the end of the story of useful ladybugs in California. In the 1920s the orchards were attacked by another scale insect, genus *Pseudococcus*. Again a ladybug, by the name of *Cryptolaemus montrouzieri*, was brought from Australia. This failed to breed under the natural conditions in western North America, so huge ladybug factories were maintained where they were bred, with careful temperature control. In 1928 alone 48 million ladybugs of this species were set free in the California orange orchards.

Ladybugs usually lay their orange eggs on the undersides of leaves, in batches of 3 to 50. A few batches are laid by one female, totaling 100 to 200 eggs, sometimes more.

LAKE AND POND

Lakes and ponds are created wherever a depression in the land surface interrupts the drainage of water off the land, causing water to collect. The most common cause of these depressions is the scouring effect of the last glaciation. Most lakes of temperate regions, including the Great Lakes of North America and the so-called thousand lakes of Finland, were created when the glaciers retreated and melted 8,000–12,000 years ago. Some older lakes were created by tectonic processes; they include the vast Lake Baikal of Siberia (25 million years old) and the Rift Valley lakes of East Africa (up to 1½ million years old).

Lake or pond?

It is possible to apply a useful distinction between the terms *lake* and *pond*. A pond is shallow, generally less than 15 feet (4.6 m) in depth. Light can reach the bottom if the water is not too cloudy, and water plants can reach the bottom of a pond with their roots. The surface of a water body defined as a pond in this strict sense can therefore be covered entirely in water lilies, for example, whereas a lake, being too deep in some parts, can be colonized by water lilies only at its edges.

However, it is not always possible to apply this simple criterion. Some huge lakes are very shallow but are clearly not ponds. For example, Lake Chad of Central Africa is no more than 13 feet (4 m) deep.

Catchment areas

Lakes usually have a clearly defined catchment area that feeds water into them by a combination of surface drainage (mainly rivers and streams) and groundwater, and an outlet that drains them. The geology of the catchment area is crucial to the kind of plants and animals that can live in a lake, because it determines the chemical nature of the water. Lakes in igneous rock basins,

Beaver Lake, Virginia, in the fall. Water lilies grow in abundance around the lake's shoreline, where the water is shallow enough for their roots to reach the bottom.

The African jacana or lily trotter,* Actophilornis africanus, *is well adapted to pond life. Its long, thin legs and toes enable it to spread its weight evenly and move easily across surface water vegetation.

for instance, are supplied with few minerals, and become oligotrophic (deficient in plant nutrients). If the drainage from the catchment area is rich in minerals, however, a lake can become full of life and very productive, or eutrophic.

Fertilizers that are washed off agricultural land can contribute to the eutrophication of lakes and ponds, and this often proves detrimental to aquatic life. Plant growth can become so vigorous that it consumes all the oxygen in the water. Water bodies become choked with algae so that only certain organisms, such as the pollution-tolerant midge larvae called bloodworms, of the phylum Annelida, can survive.

Some ponds and other small water bodies do not have a true catchment area. Instead they are filled by a single event such as a flood or a period of torrential rain.

Water quality

Distributions of terrestrial plants and animals are often controlled by climate and rainfall, but the availability of water is obviously of no concern to organisms found in lakes and ponds. For the latter the quality of the water they live in is the most important factor in their survival. Important factors include the chemical constituents of the water and whether it is well oxygenated, how much light penetrates the water surface, how quickly the lake or pond is replenished by fresh water and how quickly it empties. Two other features of lakes and ponds are crucial in influencing which animals and plants ultimately live in them: their relative isolation and whether they are long-lived or temporary.

Habitat islands

Lakes and ponds are habitat islands: isolated areas of habitat suitable for the survival of aquatic plants and animals in a relatively vast area of surrounding unsuitable habitat. This can limit the ability of organisms to colonize a newly formed lake or pond. Remote oceanic islands have few types of terrestrial plants and animals because many groups have not managed to make the journey across the sea to reach them, and a similar situation occurs with remote lakes. Nevertheless, some animals have performed considerable feats of migration to colonize very isolated stretches of water. Lake Baikal contains a large and thriving population of Baikal seals, *Phoca sibirica*, despite being thousands of miles from the nearest sea. Some experts think that the ancestors of these seals must have been helped on their long journey from the Arctic Sea by a rise in sea level and the resultant flooding over northern Siberia. Brackish conditions in the rivers might then have encouraged the seals to move upstream, until they reached Lake Baikal and evolved into a new species.

If certain groups of animals are missing from the community of a particular lake or pond, due to its isolation, the groups that have managed to make the journey are then in a position to exploit a number of vacant ecological niches. This can lead to adaptive radiation (in which an ancestral form diversifies to produce a number of specialized species) and explosive evolution (the rapid diversification of species). Both of these phenomena have occurred in the East African Rift Valley lakes, where a single family of fish, the Cichlidae, has diversified to occupy many niches usually occupied by other types of fish. The numerous species include elongated, predatory cichlids resembling pikes; smaller, bottom-feeding cichlids; cichlids that eat fish eggs; and yet more cichlids that eat mollusks.

As well as evolving to occupy most positions of the food web, cichlids seem to be particularly prone to finding specialist niches and diversifying. In Lake Tanganyika, on the border between Tanzania and the Democratic Republic of Congo (formerly Zaire) in Africa, there are about 200 species of cichlids, and about 80 percent of them are found nowhere else. The same is true of Lake Nyasa, which borders Tanzania, Malawi and Mozambique.

Become shallower with age

From the moment a lake or pond is created, it starts the process of disappearing. Sediment brought in by rivers and streams, together with dead vegetable matter, collect on the bottom. The layers of silt and rotting plants accumulate, making the body of water shallower. If the lake or pond has an outlet, this flow of water erodes a notch in the shoreline, again making the lake or pond more shallow.

In a small lake or pond, these processes, especially the accumulation of organic matter, rapidly fill up the basin. The lifetime of a pond is usually a few decades to a few centuries, after which it becomes boggy or even turns to dry land. Some ponds are seasonal: every year they are created and then evaporate.

Escaping from temporary water bodies

What do members of lake and pond ecosystems do if their habitat is isolated and temporary? Their two options are dormancy (waiting it out until the waters return) and emigration (escaping to waters elsewhere). Dormancy is an appropriate course of action if the water body in which the animals live is seasonal. Several species of desert amphibians can survive by dormancy when their temporary pools evaporate. The water-holding frog, *Cyclorana platycephalus*, of Australia, buries itself and conserves water inside a waterproof cocoon of skin and mucus. The evaporating waters of a temporary pond trigger the dormant stage of the life cycle of many organisms. The daphnia water flea, which usually reproduces asexually, starts producing males, which then fertilize the females. The eggs produced from this sexual union are specially adapted for resting for long periods until more favorable conditions return.

If the water body is unlikely to return to the same location soon, the organisms need to disperse to another lake or pond. Some rely on chance events, such as when their eggs are carried on pondweed wrapped around ducks' feet. Others reduce the element of risk with specialized adaptations. For example, the resting eggs of freshwater bryozoans (aquatic invertebrates that reproduce by budding, usually forming branched or mossy colonies), together with those of some freshwater bivalve shellfish, are armed with rings of barbs. These barbs are designed to attach to the gills of fish or to ducks' feet, by which means they might be transported to favorable habitats. Other organisms have a

Even small suburban ponds can host a wide range of wildlife. The plant life of this pond helps support frogs, newts, herons and dragonflies.

nonaquatic dispersal stage of the life cycle. The short-lived, flying adult stage of aquatic insects such as midges, mayflies and caddis-flies is an example of this strategy.

Three zones of lake life

Lakes provide at least three major habitats for organisms. The first is found around the shore of the lake where light can reach the bottom and conditions are similar to those of a pond. This is called the littoral zone. Rooted water plants, such as reeds, pondweed and water lilies, can grow here. They provide shelter, food and breeding grounds for gallinules, moorhens, rails, herons and other waterbirds, as well as for reptiles, amphibians, insects and small fish. Some fish, such as the stickleback, are always found here. Algae attach themselves to the surface of the larger plants, and grow wherever there is enough light to support their photosynthesis.

Beyond the littoral zone is the limnetic zone, the brightly illuminated, well-oxygenated surface waters of the lake. The plant life here is not rooted, and is dominated by tiny floating plants, the phytoplankton. Below the limnetic zone are the dingy depths of the profundal zone, the third lake habitat, where light does not penetrate and photosynthesis cannot take place. A major challenge for freshwater plankton is to stay afloat, and not to sink into the profundal zone. The tissues of even the smallest plankton are heavier than fresh water, so some plankton, for instance the blue-green algae *Cyanobacteria*, use gas vesicles (bladders) to control their buoyancy. Plankton without gas vesicles rely on turbulence to keep them within the surface waters. This may sound hazardous, but a process usually occurs in lakes whereby the waters in the limnetic zone, being warmer and less dense, stay on top and do not mix with the cold, dark waters below. The lake becomes stratified (layered), and water circulates within the limnetic zone, supporting its light-dependent life. Paradoxically, the productive limnetic zone becomes starved of nutrients because most of its waste falls into the profundal zone, and is not recycled. Only when the limnetic zone cools at the end of the summer do the layers mix and the nutrients recycle.

The limnetic zone is populated by fish that are powerful swimmers, such as perch and trout, and by other active swimmers such as frogs, turtles and otters. There is no cover to hide from predators away from the shore, and some prey species spend only the hours of darkness feeding in the limnetic zone, hiding in the murkier profundal or littoral zones during the day.

Most of the fish species in a lake inhabit the limnetic and littoral zones. Those that live and feed in the profundal zone are usually specialist bottom-feeders such as carp, catfish and lungfish. These fish must tolerate the dark, the cold and low oxygen levels. Most of the other organisms in the profundal zone live on or in mud. They feed on waste material falling from above, and include species such as nematode and annelid worms, bivalve mollusks and midge larvae.

The Rift Valley lakes in East Africa are home to a tremendous diversity of cichlids, which have adapted to exploit every possible food resource. Pictured is Tropheus duboisi *of Lake Tanganyika.*

LAMMERGEIER

A HUGE AND GRACEFUL vulture, the lammergeier has a wingspan of 8⅔–9⅕ feet (2.65–2.8 m). Its plumage is very distinctive: the head is whitish yellow with a black mask running from the eyes to tufts of feathers on each side of the bill, giving the lammergeier its alternative name of bearded vulture. In flight, the lammergeier is readily identified by the long, diamond-shaped tail. Its underparts are a rusty color, and the wings, tail and back are dark brown. Young lammergeiers, up to the age of 5 years, are brown all over.

The lammergeier ranges from southern Europe, including Spain, Corsica, Greece and the Balkans, east through the Middle East to northern India, Tibet, China and Mongolia. It is also found in parts of North, East and southern Africa. It lives in mountainous districts such as the sierras of Spain and the Himalayas, coming out onto the plains only to look for food. In some places, such as Tibet, the lammergeier is relatively numerous and is often found near human settlements. In most other regions it is uncommon or rare. In southern Africa, for example, there are only 50 pairs left.

Lammergeiers are carrion eaters, like all vultures, but specialize in eating bones. They drop the bones from a great height to smash them and expose the bone marrow inside.

Mountain vulture

Lammergeiers have an almost regal appearance on the wing. They are magnificent in flight, being numbered among the most skilled gliders. They soar over high mountain passes or glide close to the ground following the contours, hardly ever beating their wings. In a powerful glide lammergeiers have been timed at 80 miles per hour (130 km/h). During the breeding season they are usually found at altitudes of 3,300–14,750 feet (1,000–4,500 m). They are attracted to very steep slopes and craggy ridges, where rising currents of warm air enable them to soar more efficiently.

Scavengers of carrion

Like all vultures, lammergeiers are carrion-eaters. Other vultures such as the griffon vulture, *Gyps fulvus*, and white-backed vulture, *G. bengalensis*, are heavier and more aggressive than the lammergeiers are, and keep them from a carcass until they have finished. The lammergeiers have to make do with the leftover bones and scraps.

Lammergeiers are scavengers, although on rare occasions they take live prey. Near human habitation they eat any carrion or offal, including human corpses, and they search refuse dumps for edible scraps and manure heaps for maggots. Lammergeiers are particularly fond of bone marrow, swallowing small bones whole or dropping large bones from a height to split them. The tongue is stiff and has curved sides, apparently being used as a scoop for extracting marrow.

Inevitably, there are stories of children or lambs being seized by lammergeiers, but evidence—for the former at least—is never more than circumstantial. It has been pointed out that even if this species of vulture attacked a child, it would be unable to carry the victim away because its talons, which look so formidable, are in fact relatively weak for their size.

Inaccessible nest

Each pair of adult lammergeiers owns a large territory, which the birds rarely leave. Within this territory they build a nest of sticks, wool and other material, including bones and horns. The nest is often of considerable size, 8 feet (2.5 m) across and 2 feet (0.6 m) deep, built on inaccessible crags, on rock ledges or in caves. During the breeding season the lammergeiers, like eagles and many hawks, perform spectacular displays. They swoop and soar together, sometimes diving several hundred feet, and occasionally they roll onto their backs and grapple with their talons.

One or two eggs are laid. They are pale pink with brown and purple mottling. The female incubates most of the time, especially at night, and the eggs hatch after 55–60 days. It appears that one chick is killed shortly after hatching and only one is reared. This may be a device to limit

The lammergeier has a distinctive flight profile of very long wings and a diamond-shaped tail. The adult's orange underparts and black beard are visible only at close range.

the offspring to the number that the parents can adequately feed. If they tried to rear both, one would probably die of starvation anyway. It is worth laying two eggs because doing so means that if one cracked or was infertile, the other still stands a chance of hatching. In other words, the second egg serves as an insurance policy. If both chicks hatch, the insurance is no longer required and the second chick is killed.

The single white-coated chick is brooded by both parents for 3 weeks. After that it is often left by itself while its parents are away foraging. They bring food back in the talons or bill, or they may regurgitate it into the nest from the crop (a storage sac above the stomach). The lammergeier chick has a very wide gape and can swallow large lumps of food. It can fly in about 110 days but remains in the vicinity of the nest for some time after this.

Bone-splitters

The habit of dropping bones to split them is unique to lammergeiers. In ancient times they were called *ossifragus*, and in Spain they are still called *quebrantahuesos*, both words meaning "bonebreaker." The lammergeier swoops downwind, dropping its bone from a height of 100–200 feet (330–660 m), then immediately turns upwind and settles by the fallen bone. This maneuver, turning upwind to allow itself to control its flight carefully, enables the lammergeier to prevent other scavengers from stealing its booty. If the bone does not break, the maneuver is repeated several times. Areas of rock often become strewn with fragments of bone where a lammergeier has habitually come to drop its bones.

LAMMERGEIER

CLASS **Aves**

ORDER **Falconiformes**

FAMILY **Accipitridae**

GENUS AND SPECIES ***Gypaetus barbatus***

ALTERNATIVE NAME
Bearded vulture

WEIGHT
11–15½ lb. (5–7 kg)

LENGTH
Head to tail: 3¼–3¾ ft. (1–1.15 m); wingspan: 8⅔–9⅕ ft. (2.65–2.8 m)

DISTINCTIVE FEATURES
Huge size; strongly hooked bill; long, broad wings; long legs and feet with large talons; diamond-shaped tail. Adult: pale yellowish head with black mask and beardlike bristles (stiff feathers); orange body; dark brown upperparts and tail. Juvenile: brown all over.

DIET
Bones and meat from freshly killed mammals, tortoises and birds

BREEDING
Age at first breeding: 5 years; breeding season: eggs laid January–March (Eurasia), May–July (South Africa); number of eggs: 1 or 2; incubation period: 55–60 days; fledging period: 110 days; breeding interval: 1 year

LIFE SPAN
Up to 40 years in captivity

HABITAT
Remote mountainous areas

DISTRIBUTION
Mountains in southern and eastern Europe, Africa, Central Asia, Mongolia and China

STATUS
Generally uncommon; very rare in Europe, northwestern Africa and South Africa

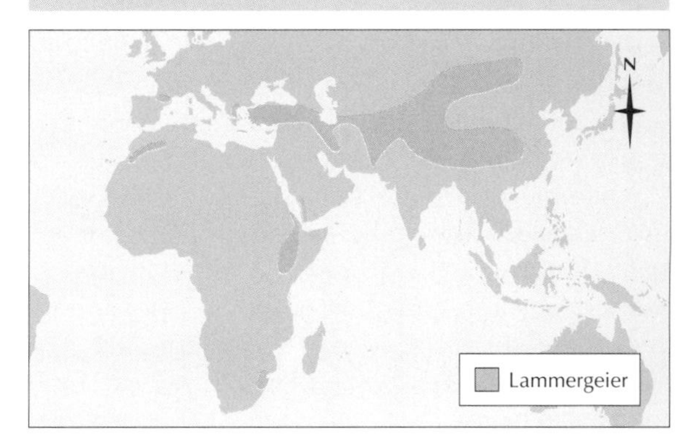

LAMPREY

Lampreys look like eels and have sometimes been called lamprey eels or lamper eels. They are, however, jawless, as are the hagfish, their nearest relatives. Today they are the most primitive living invertebrates.

There are 45 species of lampreys, both marine and freshwater. Some parasitize fish, others do not. Lampreys live in the temperate regions of the world. The sea lamprey, *Petromyzon marinus*, lives on both seaboards of the North Atlantic. Members of the genus *Lampetra* are found in Europe and Asia as well as in North America. In the Southern Hemisphere, species of the genera *Geotria* and *Mordacia* are found off the coasts of Chile, Australia and New Zealand. *Geotria* has a large fleshy bag, of unknown function, almost entirely hiding its mouth.

Pumplike gills

The lamprey has an eel-like body with a slimy, scaleless skin. Its fins are found along the center line of the body. There is a single nostril in the middle of the head, which leads behind into a blind sac. The eyes are well developed. The head ends in front in a large funnel-like mouth lined with horny teeth. Some of the teeth are mounted on the muscular tongue protruding at the base of the funnel.

Behind the head a row of small, circular gill apertures extends down each side of the body. Inside are seven pairs of gill pouches lined with blood-red gill filaments, which open into a tube that is blind at one end and opens in front into the mouth. A lamprey may breathe by taking in water through its mouth to pass across the gills. More often, because the mouth is so much used as a sucker, a lamprey breathes by contracting muscles around the gill pouches, driving the water out. As the muscles relax, water is drawn in. The pumping action seems to be helped by movements of the sinuous latticework of cartilage, the branchial basket, surrounding the gill pouches.

In most species the lamprey feeds by pressing the circular edge of its mouth against the side of a fish, which it finds by eyesight, not by smell as in the hagfish. It protrudes its tongue and punctures the fish's skin by rasping the teeth on it; the fish starts to bleed, and the blood is sucked in by the lamprey. It sucks in a few fragments of flesh as well, but it feeds more on the blood than on the flesh.

Lampreys barricade their nests

There are three species of lampreys in Europe. They are usually spoken of as the sea lamprey, the river lamprey and the brook lamprey. The last-named, smallest of all lamprey species, is confined to fresh water. Those lampreys living in the sea enter rivers to spawn. The migration begins in winter, and by spring the lampreys are in the rivers and building nests. They swim strongly and can make their way over rocks or up vertical walls, hauling themselves up with the sucker mouth. The male makes a nest by holding pebbles in his sucker mouth and moving them downstream to form a barricade. The eggs, numbering up to 300,000 in the sea lamprey, will later be laid in a depression made upstream.

Die after spawning

The females arrive later than the males and then help build the nest, the two sometimes combining forces to move large pebbles. After spawning, the adults drift downriver to die. The eggs, which are 1 millimeter in diameter, hatch two weeks later. The small, wormlike larva, or ammocoete, was once thought to be a different species. It burrows in the sand or mud, emerging at night to feed on particles of plant matter and carrion. These are strained through fleshy tentacles, or cirri, on a hoodlike mouth and passed into the gullet, where they are caught by sticky secretions on a special groove, the endostyle. In the adult the endostyle becomes the thyroid gland, the chemical controller of growth.

After 4–5½ years of larval life the ammocoete, now 4–5 inches (10–12.5 cm) long, changes into an adult lamprey. The hooded mouth becomes

***Lampreys are the most primitive vertebrates alive today. Pictured is the brook lamprey,* Lampetra planeri.**

Lampreys lack jaws and instead have circular, funnel-like mouths. They feed by latching onto other fish and sucking their blood. Pictured is the river lamprey, **Lampetra fluviatilis.**

funnel-shaped, the cirri are replaced by horny teeth, the nostril moves from the front of the snout to the top of the head and the eyes grow larger. The sea lamprey becomes silvery and goes down to the sea, as does the river lamprey, but the latter does not parasitize fish. Instead, it feeds on mollusks, crustaceans and worms. The brook lamprey, which stays in the rivers, does not feed at all as an adult.

The celebrated surfeit

It is often said that King John of England died of a surfeit of lampreys. It was, in fact, Henry I. It was King John who fined the men of Gloucester 40 marks because "they did not pay him sufficient respect in the matter of lampreys."

American history is more recent and has to do with a surfeit of lampreys in the Great Lakes. Gradually, over the years, lampreys made their way up the New York State Barge Canal and the Welland Ship Canal and became firmly established in the Great Lakes. There they ruined a commercial fishery that had been yielding a yearly catch of lake trout and other fish worth tens of millions of dollars. A major research program was established to find ways of killing off the lampreys. Weirs were built to prevent further migrations into the lakes, and the lampreys were poisoned and electrocuted. A measure of success was achieved; but a poison was then discovered that killed the larvae. The landlocked lamprey population was brought under control, and the fisheries began to recover.

LAMPREYS

CLASS **Agnatha**

ORDER **Petromyzontiformes**

FAMILY **Petromyzontidae**

GENUS ***Ichthyomyzon, Lampetra, Geotria, Caspiomyzon, Mordacia*** **and** ***Petromyzon***

SPECIES **45 species, including sea lamprey,** ***Petromyzon marinus*** **(detailed below)**

ALTERNATIVE NAMES
Nannie nine eyes; stone sucker

WEIGHT
Up to 5½ lb. (2.5 kg)

LENGTH
Up to 4 ft. (1.2 m)

DISTINCTIVE FEATURES
Eel-like but with 2 dorsal fins; skin mottled brown above, pale below; disc-shaped sucking mouth with teeth. Breeding male: dark ridge on back. Female: possesses anal fin. Juvenile: dark bluish above, white below.

DIET
Adult: sucks body fluids and flesh from living fish; also eats carrion. Riverine larva: detritus and microorganisms.

BREEDING
Adult spends 20–30 months at sea, then enters river to spawn; eggs small and not yolky, buried in spawning beds; larva metamorphosizes in fresh water

LIFE SPAN
Probably up to 11 years

HABITAT
Coastal waters and estuaries; breeds in rivers

DISTRIBUTION
Atlantic coasts of Canada, U.S. and Europe; landlocked subspecies in Great Lakes

STATUS
Uncommon

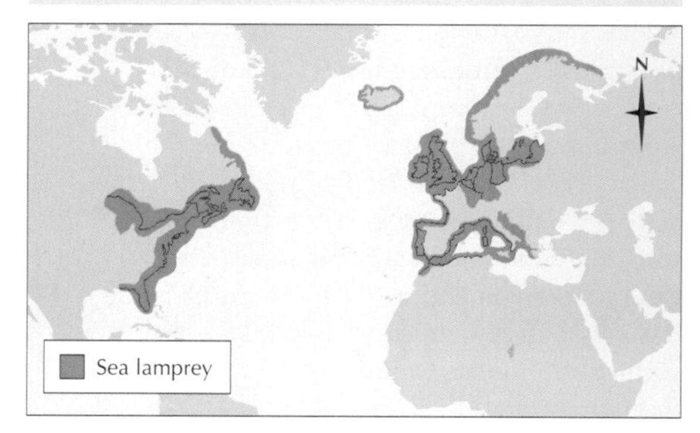

LAMPSHELL

Lampshells, which are all marine, look like clams, and until the middle of the 19th century were classified as mollusks. They are all small, up to 2 inches (5 cm) long, although some extinct forms were nearly 1 foot (30 cm) long. The shell comprises two unequal parts, or valves, which may be circular, oval or triangular. These may be made of calcium carbonate or of calcium phosphate with a horny covering. The shell is usually gray or yellow, but it may be red or orange. A major difference between a lampshell and a bivalve mollusk is that in the latter the valves can be regarded as left and right, even though one lies uppermost. In a lampshell one shell is dorsal, or uppermost, the other ventral.

Lampshells are classified in two kinds, depending on whether the two valves are hinged together or not. Usually a short, fleshy stalk protrudes at the rear of the shell. Depending on the species this may be used to anchor the animal permanently to a solid surface, such as a rock or a branching coral, or to burrow in mud. The stalk may emerge between the valves or through a special notch or hole. It is the presence of such a hole at the tip of a beaklike structure at the hind end of the larger ventral valve that causes some of the lampshells to resemble a Roman oil lamp, hence the common name.

Inside the shell most of the animal's body lies in the rear third, leaving a large space between the valves and the mantle, or layer of tissue, lining them. Within the mantle cavity lies a complex feeding organ known as the lophophore. It is covered with cilia and, although its shape varies, it usually consists of coiled arms looking like horseshoes on either side of the mouth. Along the arms are grooves and one or two rows of tentacles. The lophophore is stiffened by a skeleton of chalky rods.

There are about 300 living species of lampshells, in almost all latitudes and at most depths, although lampshells are most common in warm seas, and most live on continental slopes. In some parts of the world they may occasionally be found on the shore.

Living food pumps

Most lampshells remain permanently fixed after the larval stage. They can, however, move up and down and from side to side on the stalk. Some species, in which the stalk is rooted in sediment, can withdraw the stalk and reinsert it a little farther away. In a few species only, lacking a stalk, the shell itself is cemented to the rock. They have little need to move about since they feed on dissolved nutrients and on tiny particles, especially diatoms, that abound in the water. These are drawn in on currents set up by the thrashing of the cilia on the lophophore. Edible particles are caught on the lophophore and driven along the grooves to the mouth, which leads to a small, blind-ending stomach. Unsuitable particles, such as sand grains, are rejected and carried out again in the outgoing current. If too much silt builds up within the shell, some species can reverse the flow of the current to flush out the mantle cavity.

Mud-burrowing lampshells

One group of lampshells of the genus *Lingula* have a muscular stalk that can be shortened or lengthened and used for burrowing. There are a dozen species of these in the region of the Indian and Pacific Oceans, especially off Japan, southern Australia and New Zealand. In some places they are used for food. Their burrows, in rank, black mudflats, are vertical, 2–10 inches (5–25 cm) deep and with a slitlike opening at the top to match the animal's flattened shell. The advantage *Lingula* gains in such a habitat is that few other animals can live there. *Lingula* lies with its shell near the top of the burrow, filtering water in the

About two-thirds of the space inside the shell of a lampshell is taken up by a structure called the lophophore. This complex organ is used to catch edible food particles and pass them to the mouth.

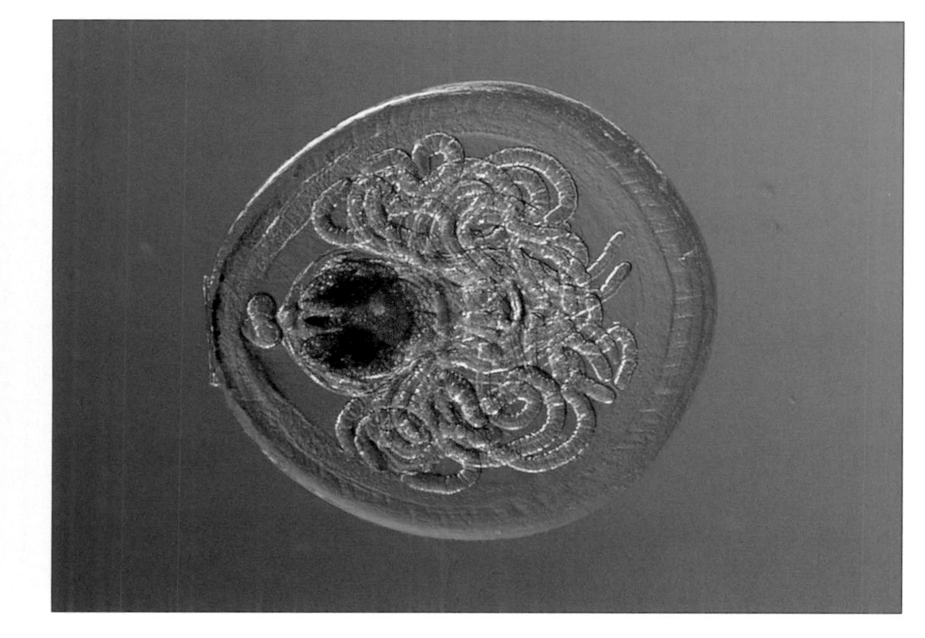

Lampshell larvae swim using cilia (minute hairlike extensions). They eventually settle on the seabed and transform into adults.

usual way. Bristles around the edges of its mantle can be bent over the opening as the shell gapes, thus keeping out the larger grains of mud and sand. The tip of the long stalk reaches to the bottom of the burrow. When the animal is disturbed the stalk contracts and is drawn down into the safety of the burrow.

Simple life history

Lampshells are usually male or female, albeit with no obvious difference between the sexes, but a few species are hermaphroditic (possessing both male and female sex organs). The eggs, heavily charged with yolk, and the sperm are passed out through the kidneys and shed into the sea, where fertilization takes place. The fertilized eggs develop into tiny larvae that swim around for a day before settling on the seabed and taking root. In some species the larvae look like miniature lampshells, with a pair of valves and a stalk that is coiled in the back of the mantle cavity. The larval lophophore is simple; its cilia are used not for feeding but for driving the larva through the water. A few lampshells brood their eggs in the mantle cavity, and some of these have a special pouch, or they may actually brood them in the kidneys.

New lamps for old

The living lampshells number around 300 species compared with 30,000 known species of lampshell fossils. In past ages they must have been as numerous as mollusks are today, and in some parts of the world whole layers of rock are made up of little else than fossil lampshells. The earliest known large animal fossil is a lampshell, and the group as a whole reached its peak during the Ordovician period, which lasted from 500 million to 440 million years ago. *Lingula* was one of the earliest; its ability to thrive in foul water is one reason why it has been able to persist almost unchanged throughout this long period of time.

There has been greater change in lampshells that more closely resemble cockles and other clams. Their shells have become ribbed or folded or ornamented with spines. Such changes may seem superficial but help geologists identify and date the rocks in which such fossils turn up.

LAMPSHELLS

PHYLUM **Brachiopoda**

ORDER (1) **Atremata**

GENUS ***Lingula* and *Crania***

ORDER (2) **Neotremata**

GENUS ***Terebratulina* and *Terebratella***

SPECIES **Total of 300 living species**

ALTERNATIVE NAME
Arm-footed lampshell

LENGTH
Across shell: up to 2 in. (5 cm)

DISTINCTIVE FEATURES
Dull grayish yellow or reddish orange shell, with unequal dorsal (upper) and ventral (lower) valves; shell is attached to substrate by pedicel (stalk), or more rarely by cementation of part or all of ventral valve

DIET
Tiny particles and dissolved nutrients

BREEDING
Have separate sexes or are hermaphroditic, depending on species; eggs develop into free-swimming larvae, which settle on ground after a few days

LIFE SPAN
Maybe up to 10 years in some species

HABITAT
Seabed, from intertidal zone to depth of about 6,500 yd. (6,000 m); most common at 165–550 yd. (150–500 m); *Lingula* burrow in sand or mud, others need firmer ground

DISTRIBUTION
Most numerous in tropical zones

STATUS
Common

LANCELET

A SEMITRANSPARENT, elongated marine animal, usually under 2½ inches (6.5 cm) long, the lancelet is shaped rather like a fish. It swims like a fish, too, by sideways undulations of its flattened body, which is pointed at each end. But it lacks the paired fins of a fish and, for other reasons, cannot qualify as a vertebrate. The various species are widely distributed in the seas throughout the world. Lancelets are found in tropical and temperate seas, generally close to shore. Originally they were given the scientific name *Amphioxus*, which means "sharp at both ends," but this has now been changed to *Branchiostoma* (for reasons which shall be given later). Amphioxus, the Anglicized form of the scientific name, is now used as an alternative common name, especially in research laboratories.

Taking evasive action

Most of the time a lancelet lies with its hind end buried in sand or gravel, the head pointing more or less vertically upward above the surface. The beating of cilia around its mouth creates a current drawing water in through the mouth and thence through a sort of sieve, known as the branchial basket, in the forepart of the body. The water passes through the sieve and out through a pore near the midriff, on the underside. When disturbed, the lancelet leaves the sand, zigzags rapidly around in the water above and then dives back into the sand a few seconds later.

Curtains of food

The branchial basket is an elongated oval with vertical slits on either side. It serves as a set of gills for taking oxygen from the water flowing through it and for capturing food. Along the floor of the basket is a groove known as the endostyle. This constantly secretes mucus that is carried up the internal sides of the branchial basket by the beating of cilia, which line the walls to form a kind of curtain. This curtain of mucus contains many tiny gaps, allowing water to permeate the gill slits and exit. Food particles, such as diatoms, are trapped by the curtain, which continues to be driven upward by the cilia until it reaches another longitudinal groove in the roof of the basket. There another set of cilia drive the mucus backward into the stomach, where it is digested.

Lopsided larvae

The male and female lancelets release their eggs and sperm into the sea to mingle. About 8 hours after each egg is fertilized a ciliated embryo has taken form. It swims about and then changes into an elongated, lopsided larva. This eventually develops into the adult. The lopsided larvae is ⅓–⅔ in. (8–16 mm) long, but sometimes it grows much larger and becomes sexually mature without changing into an adult. This process is called neoteny. The giant larvae of the lancelet were once regarded as a separate species and given the name *Amphioxides*. Depending on the species, a lancelet may live for between 1 and about 3 years.

Two lancelets of the species **Branchiostoma lanceolatum** *resting half-buried in the gravel. Their mouths are located at the tip of the exposed end of their bodies.*

A profusion of names

In 1767 a strange little animal was picked up on the coast of Cornwall in England and sent, preserved in alcohol, to the celebrated German naturalist, Peter Simon Pallas. There seems to be no record of who picked it up or why it was sent across Europe when there were many competent naturalists in Britain who might have examined it. At all events, Pallas gave it in a footnote in a book, giving a very brief description in Latin and naming the animal *Limax lanceolatus* under the impression that it was a slug.

More than half a century later, on December 21, 1831, Jonathan Couch, one of the leading English naturalists of that time, was walking along the shore near Polperro, in Cornwall, after a storm. It is the practice of some naturalists to go beachcombing after a storm to see what specimens may have been thrown ashore. Couch

Preserved specimens of* Branchiostoma californiense. *Lancelets are of great interest because, although they lack bones, sense organs and a brain, they have a nerve cord along their back. They almost, but not quite, qualify as vertebrates.

LANCELETS

PHYLUM	**Chordata**
SUBPHYLUM	**Cephalochordata**
FAMILY	**Branchiostomidae**
GENUS	***Branchiostoma***
SPECIES	***Branchiostoma californiense*; *B. lanceolatum*; others**

ALTERNATIVE NAME
Amphioxus

LENGTH
Up to 2.5 in. (6.5 cm)

DISTINCTIVE FEATURES
Elongated, narrow and flattened body; has a dorsal nerve cord but no bones or cartilage; no paired fins; distinct muscle blocks on either side of the nerve cord in an obvious chevronlike pattern; dorsal fin extends all along back; ventral fin also very long

DIET
Filters small particles, such as diatoms, from surrounding water

BREEDING
Sexes separate in most species; eggs and sperm released into water to mingle; from each egg a ciliated embryo develops into a larva, which settles on seabed after a few weeks

LIFE SPAN
Up to 3 years, depending on species

HABITAT
Sand and gravel in shallow coastal waters

DISTRIBUTION
One species or other is likely to be found in most parts of the world in suitable habitats

STATUS
Not known

apparently turned over a flat pebble lying on the sand about 50 feet (15 m) from the ebbing tide and saw a tiny tail sticking out of the sand. He dug out the rest of the animal and was able to study it in a seawater aquarium and see how active it was. Couch sent the specimen to the English naturalist William Yarrell, who described it in *A History of British Fishes* (1836) as a fish of very low organization, giving it the name *Amphioxus*. Yarrell also recognized it as the same animal that Pallas had studied. Previously, however, in 1834, the Italian naturalist Costa had published a description of the same animal collected from the shore at Naples and had given it the name *Branchiostoma lubricum*.

This brief history accounts for the changing of the animal's name. It had become generally known as *Amphioxus* because Costa's description had been overlooked and was not brought to light until 40 years ago. The international rules of nomenclature state that the first name proposed for an animal must be the one used, even if it has been overlooked for years. Accordingly, the name given by Costa had to take precedence over Yarrell's *Amphioxus*.

Invertebrate or vertebrate?

The relationship of the lancelet with the rest of the animal kingdom remains one of the most interesting features of the animal. The lancelet resembles the vertebrates in having a dorsal nerve cord lying above a stiffening rod, the notochord, and an arrangement of muscles along its tail much as in a fish. At the same time, however, it lacks a backbone, jaws, or indeed any bone, and also lacks a brain as well as the eyes and other sense organs associated with the brain, so it is not a vertebrate, yet comes near to being one. The current view is that both the lancelet and the vertebrates evolved from the same ancestors as the sea squirts or tunicates, which feed in much the same way as the lancelet but are anchored to a solid support when adult and look most unlike it. They do, however, have a free-swimming, tadpolelike larva. If this were to become sexually mature without taking on the sessile adult form, like the "*Amphioxides*" larva, we should have something resembling both the ancestral lancelet and the ancestral vertebrate.

LANCET FISH

Lancet fish are among the top predators of ocean depths, although they also occur right up to surface waters.

THE LANCET FISH TAKES the place of barracudas in the oceans' twilight zone. It has been called the "sea wolf" or "wolf fish" of the deep. There are two species of lancet fish. One occurs in the Atlantic and Pacific, the other in nearly all major oceans.

The body of a lancet fish is scabbard-shaped with a long, high, sail-like dorsal fin reaching from just behind the head almost to the adipose fin that lies in front of the well-developed tail fin. The rest of the fins are small or of only moderate size. The mouth is wide and armed with fanglike or lancet-shaped teeth. Although the largest individuals may reach 7 feet (2.15 m) long, none weighs much over 10 pounds (4.5 kg), a testimony to how long and slender the body is. The skin is scaleless.

Lone wolf of the twilight zone

The shape of the lancet fish's body and the high dorsal fin remind us of the surface-living sailfish, a swift fish. Its teeth suggest the arch-hunter, and its scaleless skin is typical of fish that are able to swallow large prey because the stomach and skin are elastic. It is easy to imagine a lancet fish flashing through the water, snapping up anything it meets, probably solitary except at the breeding season. The stomach of one lancet fish contained several octopuses, a number of crustaceans and salps, 12 young boarfish, a horse mackerel and a young lancet fish. An adult lancet fish probably has few enemies, and cannibalism, a common enough trait in such species, would act as a natural check on its numbers.

LANCET FISH

CLASS **Osteichthyes**

ORDER **Aulopiformes**

FAMILY **Alepisauridae**

GENUS ***Alepisaurus***

SPECIES **Longnose lancet fish, *A. ferox*; shortnose lancet fish, *A. brevirostris***

ALTERNATIVE NAMES
Sea wolf; wolf fish

WEIGHT
***A. ferox*: up to 10 lb. (4.5 kg)**

LENGTH
***A. ferox*: up to 7 ft. (2.15 m).**
***A. brevirostris*: up to 38 in. (96 cm).**

DISTINCTIVE FEATURES
Elongated body; scaleless, but covered with pores; no light organs; dorsal fin high and long; low anal fin; large mouth with well-developed teeth; no swim bladder

DIET
Fish, squid, octopuses, sea squirts, salps and crustaceans, including krill

BREEDING
Few details known

LIFE SPAN
Not known

HABITAT
Found from near water's surface to below 1,100 yd. (1,000 m), sometimes approaching inshore waters

DISTRIBUTION
***A. ferox*: most of North Pacific and North Atlantic, including Mediterranean.**
***A brevirostris*: most major oceans, but absent from high latitudes.**

STATUS
Little information available, but probably not threatened

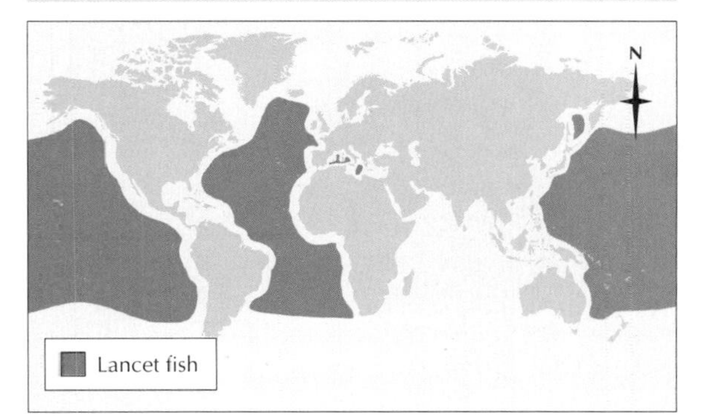

LANGUR

THERE ARE SOME 25 SPECIES of langurs. Many are colorful monkeys with crests of hair on the head and contrasting patches of naked skin, while others are more dull. Most are of slender build, 2 feet (60 cm) long with a 2½-foot (75-cm) tail, and with long fingers and toes. The Hanuman langur of India and Sri Lanka is gray washed with buff or silvery shades, often with a white head but with a black face. The male weighs 20–30 pounds (9–13.5 kg) and the female 6–15 pounds (2.7–6.8 kg). The Hanuman langur spends 50–80 percent of its day on the ground.

The purple-faced langur of Sri Lanka and southern India, black with a straggly grayish mane, is smaller than the Hanuman langur. The spectacled or dusky langur of Myanmar (Burma) and Thailand is bluish gray with white rings around the eyes and white or flesh-colored lips. The maroon langur of Borneo is bright red with a blue face, and has a crest on the back of the head. It belongs to a group of species that are small, 12–15 pounds (5.5–6.8 kg) in weight, with rounded "infantile" skulls and small faces.

The douc, a striking langur from southern Vietnam and Laos, is a speckled bluish gray with white forearms, black thighs and reddish maroon shanks, black hands and white tail. On the throat is a bright red collar bordered behind with black, while the rest of the throat is white, as are the cheeks. There is a red band on the forehead bordered with black behind, and the skin of the face is yellow. The eyes are slanted, and the nose has a curious flap overhanging each nostril. In the southern part of its range the douc is duller.

The remaining langurs have upturned noses, sometimes with flaps, like the douc. The golden snub-nosed langur is found in the bamboo forest high in the mountains of Szechwan, China. It is golden underneath and blackish brown above interspersed with long golden hairs. The upper part of the face is blue. Related forms are found in central China and northern Vietnam.

The pig-tailed langur is brownish black with an almost naked, short, curly tail. It is found only on the Mentawei Islands, a string of four deep-water islands off the western coast of Sumatra. Another species is the proboscis monkey from Borneo. In the male the long, flattened and tongue-shaped nose curves downward comically; in old males it overhangs the mouth. The nose straightens out when the animal honks its far-carrying alarm call.

One of the eight so-called true langurs is the mitered leaf monkey,* Presbytis melalophus. *It is confined to primary lowland rain forests in central Sumatra.

Different places, different habits

Troops of langurs occupy home ranges, which have a core area, or territory, into which other troops do not wander. The core contains the favorite sleeping trees of the troop. The alpha male in each troop utters a deep, resonant whoop, which seems to help in spacing out the troops. When two neighboring troops of Hanuman langurs meet, territorial wars may break out.

The behavior of Hanuman langurs shows striking regional differences that are related to habitat. Numbers range from as few as seven per square mile (less than three per sq km) in cultivated areas to as many as 220 per square mile (90 per sq km) in the forests. Troop sizes vary between 2 and 60. In cultivated areas they may become heavily dependent on crop raiding. Mixed troops contain several adult males, which are not aggressive toward one another. Both solitary males and all-male troops may be seen; these tend to avoid the mixed troops.

LANGURS

CLASS **Mammalia**

ORDER **Primates**

FAMILY **Cercopithecidae**

GENUS **True langurs, *Presbytis*; snub-nosed langurs, *Rhinopithecus*; brow-ridged langurs, *Trachypithecus*; doucs, *Pygathrix*; proboscis monkey and pig-tailed langur, *Nasalis***

SPECIES **At least 25 species, including Hanuman langur, *Presbytis entellus* (detailed below)**

WEIGHT
11–50 lb. (5–23 kg)

LENGTH
Head and body: 16–30¾ in. (41–78 cm); tail: 27–42½ in. (69–108 cm)

DISTINCTIVE FEATURES
Slender gray body; black hands and face; long black tail; head often whitish

DIET
Mainly leaves, fruits and bird eggs; also scraps in urban areas

BREEDING
Age at first breeding: 2–3 years; breeding season: March–late July; gestation period: 190–210 days; number of young: usually 1; breeding interval: 1 year

LIFE SPAN
Up to 12 years

HABITAT
Mature forests; often frequents villages and towns

DISTRIBUTION
Southern Tibet, Nepal, Sikkim, Kashmir, India, Bangladesh and Sri Lanka

STATUS
Common; estimated population: 230,000

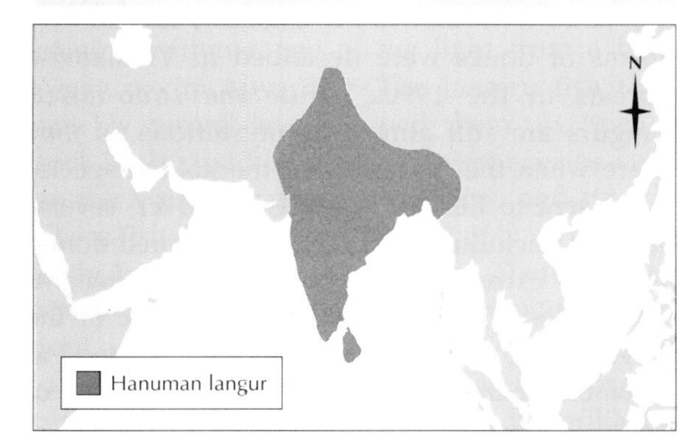

In Bangladesh and India some Hanuman langurs (female with young, above) now lead an entirely urban existence. Hanuman is the name of the Hindu monkey god.

The benevolent tyrant

The dominance order within a langur troop is not as strongly expressed as in other monkeys. It is specially obscure among the females. A higher ranking female may slap or bite a subordinate, but nothing more. A dominant male stares at a subordinate, slaps the ground, grimaces, crouches, then suddenly stands again, grunting. Then he tosses his head, lunges at and chases the subordinate, hitting and even biting him. However, the alpha male is generally good-natured. A subordinate may approach him and "present" his rump in a sign of submission, then lie down and groom the alpha male. At the end of the session the alpha male dismisses him with a hand gesture.

Langurs rise just before dawn and often come to the ground when the sun rises. During the day they may wander up to 2 miles (3.2 km). They feed in the morning and late afternoon, and for 2–4 hours over midday they rest and groom each other. The stomach has an extended upper

Leafcutter ants perform feats of which no human weight lifter is capable. They often carry leaf fragments larger than themselves. The fragments are usually two to four times the ants' own body weight.

rejected as a pellet. When the *Atta* queen leaves her nest, she always carries in her infrabuccal pocket a pellet made up of the hyphae, or living threads, of the fungus. As soon as she settles down in the crevice in which she will found her nest, she ejects the pellet and then carefully cherishes and cultivates it, using her own excrement as fertilizer. The fungus grows and provides the food without which no *Atta* colony can exist.

Division of labor

Within each colony there is a division of labor. The largest workers, the maximae, defend the nest, and the media workers collect the leaves. Smaller workers then snip the leaves into tinier pieces, roughly 1 millimeter across. Even smaller ants, the minimae, then chew the chunks into the moist pellets that are added to the fungus beds. They also scuttle through the tiny channels in the fungus, occasionally ripping out chunks to feed to larger workers.

Minimae have other duties, too. When the mediae are out harvesting, they are exposed to attacks by a small fly, genus *Apocephalus*, which tries to settle on an ant's neck. If it succeeds, it lays an egg, which hatches into a parasitic larva that eats out the ant's brain. Normally a media can snap with her jaws to fend off the fly, but she cannot do so when working, so a tiny minima acts as her "bodyguard." It keeps guard when the media is busy cutting a leaf, and then rides on her load as she returns home, perched near the top with its jaws spread menacingly. It darts and snaps at any fly that tries to settle.

LEAFCUTTER ANTS

PHYLUM	**Arthropoda**
CLASS	**Insecta**
ORDER	**Hymenoptera**
FAMILY	**Formicidae**
GENUS	***Atta*; others**
SPECIES	**At least 200 in genus *Atta***

ALTERNATIVE NAME
Parasol ant

LENGTH
⅒–¾ in. (2.5–2 cm); several castes of considerable variation in size

DISTINCTIVE FEATURES
Spines on body; long legs; rusty brown in color. Queen: large; has wings. Minima worker: tiny; small head; chews leaves in nest. Media worker: medium-sized; collects leaves. Maxima worker: very large; strong jaws; defends nest.

DIET
Fungus cultivated by the ants in their nest

BREEDING
Fertilized female lands, shears off wings and tunnels out first nursery chamber of new colony, of which she will be queen; she uses a sample of fungus to start new garden, then lays an initial batch of 3 to 6 eggs

LIFE SPAN
Queen: more than 15 years

HABITAT
Sandy, well-drained soils in forest and wooded areas

DISTRIBUTION
Southern Texas, Louisiana and Mexico south to Brazil and northern Argentina

STATUS
Locally abundant

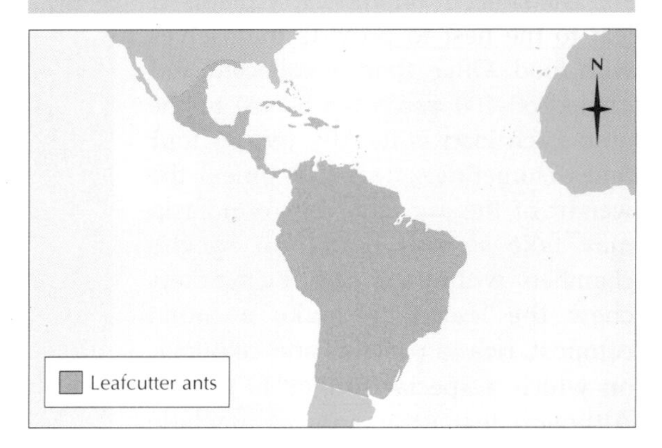

LEAFCUTTER BEE

A NUMBER OF BEES ARE known by this name from their habit of cutting neat pieces out of leaves to line the cells of their nesting burrows. They belong to a large group of solitary bees that have pollen baskets, receptacles formed of bristles for collecting pollen, on the underside of the abdomen. In honeybees and bumblebees the pollen baskets are on the hind legs. Leafcutter bees also have large heads and well-developed jaws.

A well-known leafcutter bee is *Megachile centuncularis*, which burrows in decaying wood. Other species use nest chambers already made by other wood-boring insects, and a few burrow in soil or sand.

Habits and life cycle

The smoothly rounded cutouts in the leaves of rose bushes and other plants, which so often mystify and infuriate gardeners, are a sure sign of the activities of leafcutter bees. They use the sections they snip out of leaves to line their burrows and to separate the cells. The breeding habits of the leafcutters are basically the same as those of other solitary bees. After mating, the female bee digs a burrow or finds one ready-made, and builds a row of compartments, known as cells. In each of these she stores a mixture of honey and pollen and then lays an egg on it. This food is enough for the whole development of the larva. Oval pieces of leaf are cut for the walls of the burrow and arranged so as to overlap. The cells are divided, and the last cell is capped by circular pieces of leaf. The final result is a series of cells in the burrow, looking like small cigar stubs laid end to end.

In species that produce only one generation per year, the bees hatch from their pupae in the autumn but do not come out as perfect insects until the spring. When they do so, the one in the outermost cell, and therefore the youngest by a narrow margin, comes out first, and the rest follow in succession. Some unknown mechanism restrains each bee from attempting to burrow out through an occupied cell. The outer cells almost always contain male bees, the inner ones females.

Leafcutter bees can be studied by drilling holes in logs or dead trees or by putting out lengths of hollow plant stems. Sometimes they will accept accommodation provided in this way, so their nesting can be watched.

Bees and alfalfa seed

Lucerne or alfalfa is one of the main forage crops in the United States, but there has long been a problem about its seed production. When a bee visits a lucerne flower, it forces apart the petals that enclose the anthers and stigma and form what is known as the keel of the flower. This releases a sort of trap by which the reproductive

A female leafcutter bee about to land at her burrow with a section of rose leaf. She will collect many similar pieces, using them to line the burrow and separate the cells inside.

Leafcutter bees have unusually large heads and well-developed jaws. In parts of the United States the bees are bred to pollinate fields of alfalfa.

LEAFCUTTER BEES

PHYLUM **Arthropoda**

CLASS **Insecta**

ORDER **Hymenoptera**

FAMILY **Megachilidae**

GENUS ***Megachile***

SPECIES **About 1,500, including *M. centuncularis* and *M. rotundata*; 120 species in North America**

ALTERNATIVE NAME
Alfalfa bee

LENGTH
About ⅖–⅗ in. (1–1.5 cm)

DISTINCTIVE FEATURES
Resembles honeybee but is darker in color, with pale bands on abdomen, much larger head and well-developed jaws

DIET
Flower nectar and pollen

BREEDING
Female constructs nest within plant stem and makes nest cells using cut pieces of leaves; number of eggs: about 35 to 40 in female's lifetime; larval period: adult bees emerge in following year

LIFE SPAN
Adult female: up to 2 months

HABITAT
Wide variety of habitats; nests in soft, rotted wood in stems of large, pithy plants such as roses

DISTRIBUTION
Almost worldwide; in Americas, ranges from Alaska south to Tierra del Fuego

STATUS
Common

parts of the flower spring up and hit the bee on the head. A honeybee learns to avoid this rude reception by maneuvering its tongue straight to the nectar without parting the enclosing petals. Its caution results in anthers and stigmas being avoided altogether; the flowers are not pollinated, and no seed is produced.

Certain solitary bees are somewhat less sensitive and ignore their rough reception. They are, however, not nearly numerous enough to pollinate effectively the vast fields of alfalfa that are grown for seed. Some years ago, Professor W. P. Stephen at Oregon State University pioneered the mass breeding of two species of solitary bees artificially in alfalfa-growing districts, increasing the yield of seed as much as 10-fold. One of the species, the alkali bee, *Nomia melanderi*, is of no concern here, but the other is a small species of leafcutter, apparently an accidental introduction from Europe. It is a species that nests in ready-made burrows and is sociable, favoring densely packed burrows. After experimenting with holes drilled in wood, Stephen found that paper drinking straws, packed and glued into their original containers, were readily used by the bees. The straws were closely packed, but each female appeared to have no difficulty in locating her own nest among the 200-odd identical circles that faced her as she flew in toward a full container. In many cases almost all the straws were occupied. The bees favored straws 5 millimeters in diameter and 4 inches (10 cm) long.

The culturing of this bee is now well established, and the seed producers make boxlike homes for the bees, roofed and protected against birds with wire netting, and packed with large numbers of drinking straws in their containers. One such domicile, if well stocked with bees, is sufficient for about 5 acres (2 ha) of crops, and the beauty of this method is that the bees can be moved from place to place as required. One species of leafcutter bee cultivated for alfalfa pollination in Colorado is *Megachile rotundata*. The bees are reared in predrilled "bee boards" that they use for nest building. At the end of the season, the nest cells with developing bees are collected and carefully stored, to be released the following season when the alfalfa blooms.

LEAF FISH

IN 1840 AN ODD LITTLE FISH was added to the collections in the Vienna Museum. It came from the northern part of South America and was given the scientific name *Monocirrhus polyacanthus*. The local South Americans call it the leaf fish. Between 1822 and 1954 10 species of leaf fish were discovered: two in West Africa, six in India and southern Asia and two from Central and South America. In 1996 came the description of a new species of leaf fish from the Mekong Basin in Thailand: *Nandus oxyrhynchus*.

The largest leaf fish is the Malaysian leaf fish, *Pristolepis fasciata*, which measures up to 8 inches (20 cm) in length. The body is flattened from side to side, roughly oval in outline, with a pointed snout and rounded tail fin. The dorsal fin is spiny with a soft-rayed portion at the rear, and the anal fin is similar.

Judging by their discontinuous distribution, leaf fish are primitive. They may have evolved as a group in fresh water many millions of years ago, before the continents drifted apart, although this has not been proved, there being no fossil record for the family. A similar species, the blue perch, *Badis badis*, of India, is subject to great variation in its bright colors. It may be brown with a black or red chain pattern, pale blue fins edged with pink and yellowish markings on the body, or it may be dark brown with black markings, blue fins and buff markings on the body, or even a pale flesh color with dark markings, blue fins, red blotches and fawn patches.

Remarkable camouflage

Leaf fish generally live in still or sluggish waters, keeping to the shadows. They move little, except when taking prey, and they tend to hold a position, often with the body pointing obliquely down. The fins may be kept folded or extended to show the saw-edge of the spiny dorsal and anal fins. At all times they look like leaves; the resemblance is enhanced by the mottled color. In the Amazon leaf fish, and to some extent in other species, the eye is obscured by dark lines that radiate from it and mask its outline. The male of this species also has a barbel on the chin that looks like a leaf stalk, and in most species the tail fin and the rear soft-rayed portions of the dorsal and anal fins are so delicate that they seem to vanish, making the body look less like a fish than a dead and sodden leaf.

Deadly leaves

Not only does it look like a dead leaf, the leaf fish also mimics one when hunting. It drifts slowly toward a fish, then suddenly shoots out an extensible and capacious mouth to claim a victim. The leaf fish is a voracious predator, able to swallow another fish up to three-quarters its own size. Normally it eats its own weight of food in a day. Although leaf fish live well in aquaria, they should not be put in mixed tanks; they also tend to be cannibalistic, especially when young.

Cleaning up before spawning

In all species there is no great difference between the sexes, but breeding behavior varies within the family. The Amazon leaf fish has no courtship display, but the two partners clean the surface of a broad leaf of a water plant, or of a stone, and the eggs are laid on this. Only a few eggs are laid on the cleaned surface, and these hatch in 3 or 4 days. At first, the male takes care of the eggs, fanning them with his fins. The near-transparent fry (young) feed on water fleas. At about a month old, the members of a brood, having grown at different rates, become destructive, the larger eating the smaller. At about this time also, white spots cover their bodies. These later fade. The spots may serve as camouflage.

Most species of leaf fish spawn in pits in the sand or gravel, although one makes a bubble nest. At least one species lays its eggs in small crevices among stones.

Mottled coloration, a flattened body and saw-edged dorsal and anal fins complete the near-perfect disguise of the Amazon leaf fish.

Leaf fish hunt by slowly drifting toward another fish, then suddenly shooting out their large, extensible mouth to claim the surprised victim.

Making her presence felt

When a pair of fish looks so closely like dead leaves that either can drift undetected toward its prey, it may be wondered how they identify each other as fish, let alone as potential mates. Clues are found in the courtship of the blue perch. The male chooses a dark cavity among rocks or stones. He wriggles over the sandy floor to scoop out a depression. If the floor is a rock surface, he uses the same action to clear it of algae. A female draws near the entrance and the male stops work, as if sensing her. He drives her away and returns to his task. She comes back and is again driven off. This goes on repeatedly until the male has completed his preparations. When the female now returns, she tilts her body toward him, presenting to him her swollen abdomen. The male turns black but the female becomes pale as she enters the nesting cavity. They swim around each other, and for a brief moment they meet mouth to mouth in a kissing action. He then wraps himself around her middle and the two drop to the floor of the cavity, where she sheds her eggs. Sometimes she scatters her eggs as the two hang suspended in an embrace a little way up from the bottom; or she may turn upside down and lay her eggs on the roof.

In both sexes other color changes will have taken place during courtship. These are probably visual signals, as is the female's action of turning her swollen abdomen toward him. In tests water has been taken from an aquarium in which there is a female ready to breed. When this is poured into a tank in which a lone male is preparing a breeding cavity, he behaves as though there is a female in the tank engaging in a courtship ritual. It seems therefore that as she comes closer to spawning condition, she gives off a chemical, possibly a pheromone (a sexual scent), into the water to signal her readiness to the male.

LEAF FISH

CLASS **Osteichthyes**

ORDER **Perciformes**

FAMILY **Nandidae**

GENUS ***Afronandus, Nandus, Monocirrhus, Badis, Polycentropsis, Polycentrus* and *Pristolepis***

SPECIES **11 species, including *Monocirrhus polyacanthus* (detailed below)**

ALTERNATIVE NAMES
Amazon leaf fish; barbeled leaf fish; South American leaf fish

LENGTH
Up to 3⅛ in. (8 cm)

DISTINCTIVE FEATURES
Roughly oval body, flattened from side to side; pointed snout; rounded tail fin; spiny dorsal and anal fins with soft-rayed portion at rear; coloration and chin barbel increase species' resemblance to a leaf with a stalk

DIET
Other fish

BREEDING
Eggs laid on a stone or leaf of a water plant; number of eggs: several; hatching period: 3–4 days; breeding interval: 1 year

LIFE SPAN
Not known

HABITAT
Depths of still or sluggish tropical freshwater systems

DISTRIBUTION
***Monocirrhus polyacanthus*: tropical South America, from Guyana south to central Brazil and northern Bolivia**

STATUS
Locally common

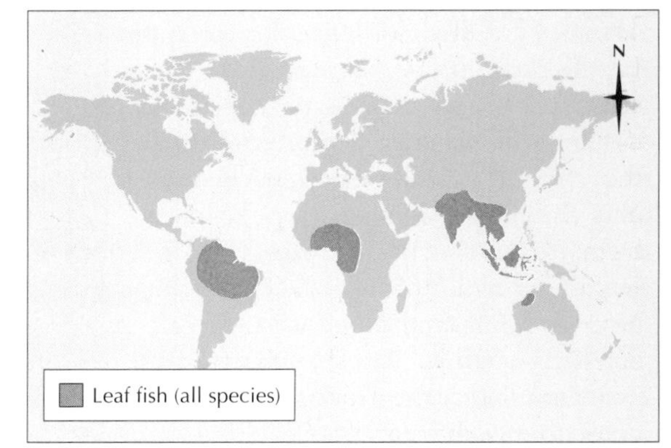

LEATHERBACK TURTLE

A female leatherback turtle coming ashore to lay her eggs. The upper shells of leatherbacks are made of hundreds of bony plates covered with extremely tough, leathery skin.

THE LEATHERBACK OR leathery turtle, sometimes known as the luth, is the largest sea turtle. It differs from other turtles in the structure of its shell. The upper shell, or carapace, is made up of hundreds of irregular bony plates covered with a leathery skin instead of the characteristic plates. There are seven ridges, which may be notched, extending down the back, and five on the lower shell or plastron. Leatherbacks are dark brown or black with spots of yellow or white on the throat and flippers of young specimens. They grow to a maximum length of 9 feet (2.7 m), of which the shell constitutes up to 7⅞ feet (2.4 m), and may weigh up to 1,800 pounds (835 kg). The foreflippers are enormous; leatherbacks 7 feet (2.1 m) long may have flippers spanning 9 feet (2.7 m).

Tropical wanderers

Leatherback turtles are found mostly in tropical seas; they spend more time in deep waters than do the other six species of marine turtles. During the summer months they may wander considerable distances away from the Tropics; they have been found as far north as Alaska, Labrador, Iceland, Scotland, Norway and Japan in the Northern Hemisphere, and as far south as Argentina, Chile, most of Australia (excluding Tasmania) and the Cape of Good Hope. They enter the Mediterranean Sea from time to time, but do not breed there. It is thought that these movements reflect the migration of jellyfish, which are the mainstay of their diet.

Survival in cool water is aided by "countercurrent heat exchangers" in the flippers. The blood vessels are arranged in such a way that some of the heat that has been produced in the powerful muscles used for swimming is transferred from arterial blood being carried from the body to the flippers to venous blood that is returning to the body. Furthermore, leatherback turtles have very oily flesh, and this may help to insulate them.

A soft diet

The stomach contents of leatherbacks show that they feed on jellyfish, pteropods (planktonic sea snails), salps and other soft-bodied, slow-moving animals, including the amphipods and other creatures that live inside the bodies of jellyfish and salps. Leatherbacks have been seen congregating in shoals of jellyfish, snapping with their horny beaks at the prey. The horny spines in the mouth and throat, 2–3 inches (5–7.5 cm) long, are probably a great help in holding slippery food. In recent years marine pollution has caused problems for leatherback turtles, which sometimes mistake floating polythene bags and other plastic garbage for jellyfish and try to eat them. The plastic is indigestible and may kill the turtles.

Most hatchlings perish. On beaches they fall prey to crabs, monitor lizards, vultures, small cats, raccoons, gulls and frigate birds. If they manage to reach the sea, the babies must then face predatory fish and squid.

LEATHERBACK TURTLE

CLASS **Reptilia**

ORDER **Testudines**

FAMILY **Dermochelyidae**

GENUS AND SPECIES ***Dermochelys coriacea***

ALTERNATIVE NAMES
Leathery turtle; luth

WEIGHT
Up to 1,800 lb. (835 kg); usually less than 1,200 lb. (545 kg)

LENGTH
Largest on record had a shell 7⅚ ft. (2.4 m) long; shell usually up to 6 ft. (1.8 m)

DISTINCTIVE FEATURES
Largest marine turtle in world; outer part of carapace (shell) made of tough, leathery skin rather than hard scales

DIET
Mainly jellyfish; also salps and pteropods (planktonic sea snails)

BREEDING
Age at first breeding: probably 8–15 years; breeding season: April–October (Caribbean), December–April (Indian Ocean); October–March (Eastern Pacific); number of eggs: 60 to 150, laid in 4 or 5 batches; hatching period: usually about 50 days; breeding interval: 2–3 years

LIFE SPAN
Probably at least 30 years

HABITAT
Tropical oceans; wanders into subtropical and temperate waters in summer

DISTRIBUTION
Worldwide except Arctic and Antarctic

STATUS
Endangered; estimated total population: 150,000 to 400,000 adults

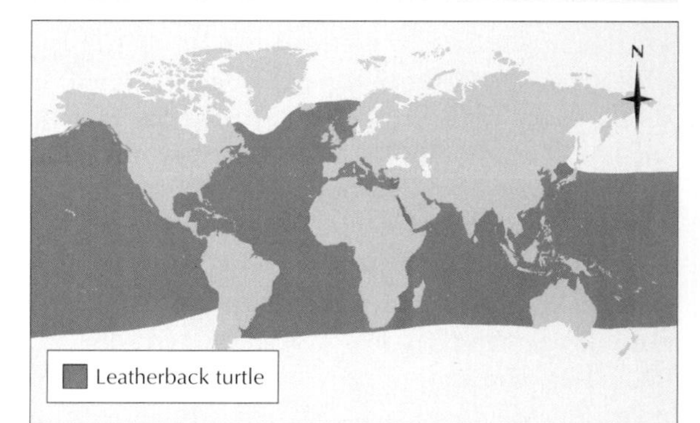

Life is a beach

Breeding takes place on beaches in many parts of the world, including the east Florida coast and at several sites around the Gulf of Mexico. Other nesting beaches are found in Central and South America, the Caribbean, the Atlantic coast of Africa, and at many sites in the Pacific. In total, there are more than 70 nesting beaches for leatherback turtles, many of them shared with other sea turtle species.

Females come ashore in small bands to lay their eggs, usually late at night. They come straight up the shore to dry sand, stop, and then start to dig the nest. They do not select the nest site by digging exploratory pits and testing the sand, as do green turtles. Each female digs with all four flippers, working rhythmically, until she is hidden in the sand. She then digs the egg pit, scooping with her hind flippers until she has dug as deep as she can reach. About 60 to 150 eggs, 2–2¼ inches (5–5.7 cm) across, are laid; then she fills the nest with sand and packs it down. Finally she masks the position of the nest by plowing about and scattering sand, and then makes her way back to the sea.

Each female lays about about four times in one season. The eggs hatch in 7 weeks, and the babies hatch together and rush down the sandy shore to the water. Atlantic leatherback turtles usually breed between April and November; the timing is more variable in the Pacific Ocean.

LECHWE

Lechwe are antelopes closely related to both waterbuck and kob. There are two species: the lechwe, *Kobus leche*, and the Nile lechwe, *K. megaceros*. Both have a longish, rough coat but no mane, and the long, slender, lyre-shaped horns, in males only, are curved twice. They have no face glands, and the glands of the groin are rudimentary. The Nile lechwe, which is the smaller of the two, lives in the Sudd swamps flanking the White Nile and the Bahr-el-Ghazal. It stands 33–40 inches (0.85–1 m) high; the adult male is a reddish black-brown with a white pattern on the head, extending from in front of the eyes to behind the base of the horns, and then down the nape of the neck to a white patch on the withers. The chin and upper lip, the middle of the belly and the inner surfaces of the hind legs are white, and there is a broad white band above the hooves. The female and young are yellowish brown, with weakly defined white areas on the head and no white on the neck or shoulders. The Nile lechwe is a rare species today and has been little studied.

The lechwe, *Kobus leche*, of the Zambezi region stands up to 43 inches (1.1 m) high with long, coarse hair. The three subspecies are so different in appearance that each warrants a description. The most widespread is the red lechwe, *K. l. leche*, of the upper Zambezi and its tributaries, western Zambia, southeastern Angola, the Ngami swamps of Botswana and the Caprivi Strip of southwestern Africa. A pale tawny red above, it is white below, from belly to chin; the shins and the hind shanks are also white. The Kafue lechwe, *K. l. kafuensis*, is found only on the marshes of the middle Kafue River, a tributary of the Zambezi. In this subspecies the spreading horns reach 32 inches (81 cm) in length, and the black line up the foreleg expands on the shoulder to form a bold patch. In the black lechwe, *K. l. smithemani*, the black has spread still farther, so that it covers the entire head, upperparts and fronts of the limbs. The black lechwe is found from Lake Bangweulu to Lake Mweru and over the Zambia border into the Democratic Republic of the Congo (Zaire). It is the smallest subspecies. A male red or Kafue lechwe weighs 220–260 pounds (100–118 kg), whereas a female weighs 165–185 pounds (75–84 kg), while a male black lechwe weighs only 150–200 pounds (68–91 kg).

Sustained by floods

The vast marshes known as the Kafue Flats, 140 miles (225 km) long, 10–30 miles (30–48 km) wide and covering 2,500 square miles (6,475 sq km) of the Zambesi Basin, are home to the best-known of the lechwe. In the rainy season, from November to March, the flats are flooded; from April to October they are dry, with water virtually confined to the river itself. Accordingly, the

Mock fights between male lechwe occur at all seasons, but at breeding time the fighting is in earnest.

A male red lechwe in the Okavango Delta, Botswana. Lechwe live in swamps, marshes and reed beds, but are forced to take refuge on drier land during the annual floods.

lechwe lives on the margins of the river and in the shallows, feeding on grasses both in the water and on land. The herds are very loose and flexible, and the sexes tend to maintain separate herds. When the floods are at their height, at the end of the rainy season, the lechwe are confined to a narrow belt extending ¾ mile (1.2 km) inland from the water's edge. At this time there is not much grazing to be had in the water, and they feed mainly on dry pastures that have long been ripe and so have little food value. The lechwe are generally in poor condition. In June and July the floods recede, revealing hundreds of square miles of grazing, and their condition improves dramatically. Later in the year, when the rains begin again, the storms drive the other animals, such as zebra and wildebeest, off the flats, and the lechwe have the area to themselves.

Survival of the very fittest

The mating season, or rut, takes place from late October until early January, but it continues sporadically until the end of the rains. Mock fights between the bucks occur at all seasons, but at breeding time fighting is in deadly earnest. Wounds, injured limbs and broken horns are common. Among black lechwe the fights are sometimes fatal. The bucks make staccato grunts that are audible ¼ mile (0.4 km) away.

No territories are formed; the males merely gather harems. Of the one-year-old does 40 percent breed, and nearly all breed in subsequent

LECHWE

CLASS **Mammalia**

ORDER **Artiodactyla**

FAMILY **Bovidae**

GENUS AND SPECIES **Lechwe, *Kobus leche*; Nile lechwe, *K. megaceros***

ALTERNATIVE NAMES
***K. leche leche*: red lechwe; *K. l. kafuensis*: Kafue lechwe; *K. l. smithemani*: black lechwe. *K. megaceros*: Mrs Grey's lechwe.**

WEIGHT
110–260 lb. (50–118 kg)

LENGTH
Head and body: male 5¼–6 ft. (1.6–1.8 m), female 4¼–5½ ft. (1.3–1.7 m); shoulder height: male 2¾–3⅔ ft. (0.85–1.1 m), female 2¾–3 ft. (0.85–0.9 m)

DISTINCTIVE FEATURES
Long-haired coat, glossy brown above and white below; white eye-rings and bib; black leg stripes (vary according to subspecies); splayed hooves; lyre-shaped horns (male only)

DIET
Grasses

BREEDING
Age at first breeding: 2–6 years (male), 1–2 years (female); breeding season: all year; gestation period: about 230 days; number of young: usually 2; breeding interval: 1 year

LIFE SPAN
Up to 21 years

HABITAT
Reed beds, marshes and bush; near water

DISTRIBUTION
Nile lechwe: near White Nile, Sudan. Lechwe: south-central Africa.

STATUS
Lechwe: common in some places, but scarce in others. Nile lechwe: near threatened.

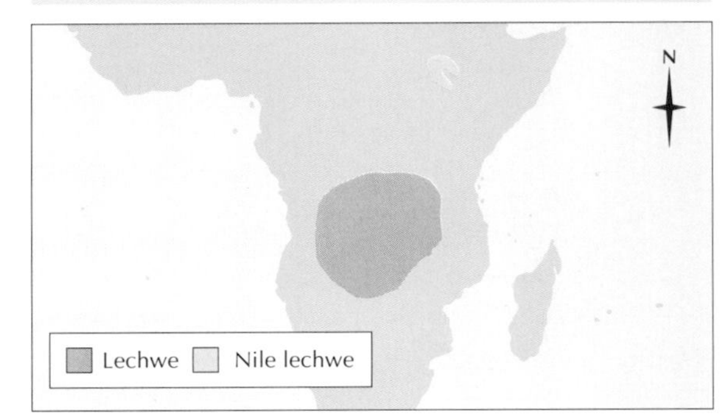

years. Gestation lasts 7–8 months. The lambs are dropped from May to December with a peak from mid-July to late August. The does leave their herd singly or in groups and give birth in patches of tall grass, often on high ground surrounded by floodwater. The lambs stay hidden for 3–4 weeks before rejoining the herds with their mothers. Half of the lambs die before weaning, which occurs at 3 to 4 months; in the dry season many lambs of the yearlings die of a warble infestation. Thus, selection in the lechwe is very severe indeed and those that survive are, in strictest Darwinian terms, "the fittest."

Males are full-sized at 4 years, females at 3 years. The males' horns appear first at 5 to 6 months; at a year they are 6 inches (15 cm) long; at 2 years, 15 inches (38 cm); at 3 years, 23 inches (58 cm); at 4 years they average 32 inches (81 cm). The record length is 36¼ inches (92 cm). The black leg stripe appears at 3–4 years of age.

Lechwe are preyed on by hyenas, crocodiles, cheetahs and African hunting dogs, in descending order of severity. Lions also prey on them in some areas. Pythons, leopards and large eagles take a toll on the young lechwe.

Slaughter in the marshes

Lechwe are of immense importance in the ecology of the marshes. They provide vast quantities of dung, which makes the area fertile and able to support large numbers of animals, not only lechwe, but also wildebeest, zebras and domestic cattle. Lechwe dung also improves the fertility of the water, which makes for a prosperous fishery and abundant waterfowl.

The amount of protein on the hoof, on the wing and in the water on the Kafue Flats is very great. In the 1930s it was estimated there were 250,000 red lechwe on the Flats, probably an overestimation, with the true figure likely to be nearer 160,000. In 1960 there were only 25,000. Excessive hunting was the main reason for this dramatic decline in numbers.

The black lechwe has also been severely overhunted by local people. The population of this subspecies fell from 1 million in 1900 to 150,000 in 1934, to 16,000 in 1959, though it seems not to have dropped too much since then. The Zambia Wildlife Society has pressed for the creation of a Black Lechwe Reserve to save this subspecies from extinction.

The lechwe plays an important part in the ecology of swamps. It is one of the habitat's main grazing animals and its dung enables lush plant growth.

LEGLESS LIZARD

THERE ARE MANY KINDS of legless lizards in the world, such as the slowworm, *Anguis fragilis*, of western Europe, and the glass lizards of the widespread genus *Ophisaurus*.

Lying low in California

One kind of legless lizard lives in parts of California and is often called the California legless lizard. It looks rather like a small snake. The presence of moveable eyelids, however, demonstrates that it belongs to the lizard suborder, Sauria, and not the snake suborder, Serpentes, of the order Squamata. The California legless lizard may grow to 10 inches (25 cm) in length. It has smooth, shiny scales, and the back is usually a silvery gray or, more occasionally, beige in color. A black line extends from behind the neck along the center of the back, and other dark longitudinal lines may be present. The underside is yellow. Individuals from the Monterey Bay region may be dark brown or even black on the dorsal surface, but they remain yellow beneath.

California legless lizards are almost confined to their namesake state; their range stretches from the San Joaquin River to the extreme south of the state and into a very small extent of Baja California. They do not range more than about 150 miles (240 km) inland. A closely related species is found on the island of Geronimo and the adjacent mainland in Baja California; it is often called the Geronimo legless lizard.

Burrows to avoid heat

Legless lizards spend much of their time burrowing. They feed on insects, worms and other arthropods in the soil, such as wood lice and mites, centipedes and millipedes, which they find by smell and touch.

The depth at which the lizards are found varies with the time of year. During winter and spring, when the soil is not heated to any great depth, they are 6–12 inches (15–30 cm) down. As summer progresses, they go deeper, to 3 feet (0.9 m) in early summer and 4–5 feet (1.2–1.5 m) in midsummer. Several zoologists have looked into this. They have taken the temperature of the lizard immediately on capturing it, the temperature of the soil at the depth of the lizard's burrow, and the air temperature. There is a fairly steady correlation between these, with the lizard's body temperature about 2° F (1° C) above that of the soil and the temperature of the soil about 4° F (2° C) above that of the air at the surface. The lizards evidently follow a "temperature gradient," moving down as the soil heats up in summer and up as the soil cools in the winter of subtropical California.

Legless lizards are sometimes found under logs and boulders, where temperatures are more stable. They rise to the surface at night, and there are records of legless lizards crawling across streets, in Bakersfield, Kern County, moving from a piece of uncultivated ground. One had traveled 500 yards (460 m) when picked up.

The California legless lizard may resemble a snake, but its moveable eyelids and lack of a forked tongue confirm its identity as a lizard. Pictured is the dark subspecies,* Anniella pulchra nigra, *from the Monterey Bay region.

CALIFORNIA LEGLESS LIZARD

CLASS **Reptilia**

ORDER **Squamata**

SUBORDER **Sauria**

FAMILY **Anniellidae and Anguidae**

GENUS AND SPECIES **California legless lizard, *Anniella pulchra* (detailed below); Geronimo legless lizard, *A. geronimensis*, various others**

LENGTH
Up to 10 in. (25 cm)

DISTINCTIVE FEATURES
Slender-bodied legless lizard, distinguished from snakes by its moveable eyelids and lack of forked tongue; back usually silvery gray in color, with black stripe extending down center

DIET
Earthworms, insects and other invertebrates living in soil and leaf litter

BREEDING
Ovoviviparous (producing live young). Age at first breeding: 3 years; breeding season: summer.

LIFE SPAN
Not known

HABITAT
Damp soil or leaf litter in variety of habitats; often associated with bush lupins

DISTRIBUTION
From San Joaquin River (Central Valley, California) to northern tip of Baja California, up to 150 mi. (240 km) inland from coast

STATUS
Patchy distribution within range; population estimates uncertain

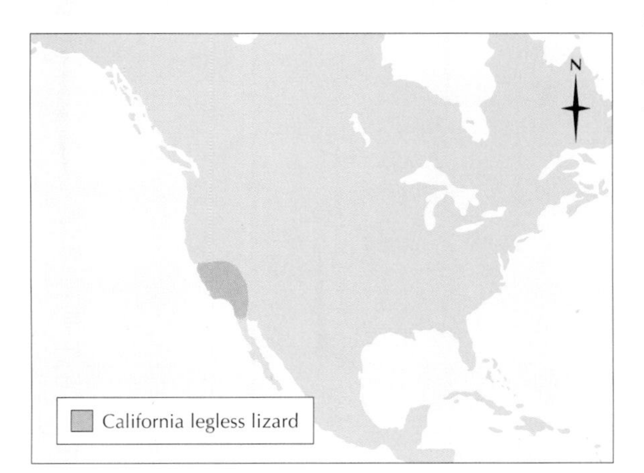

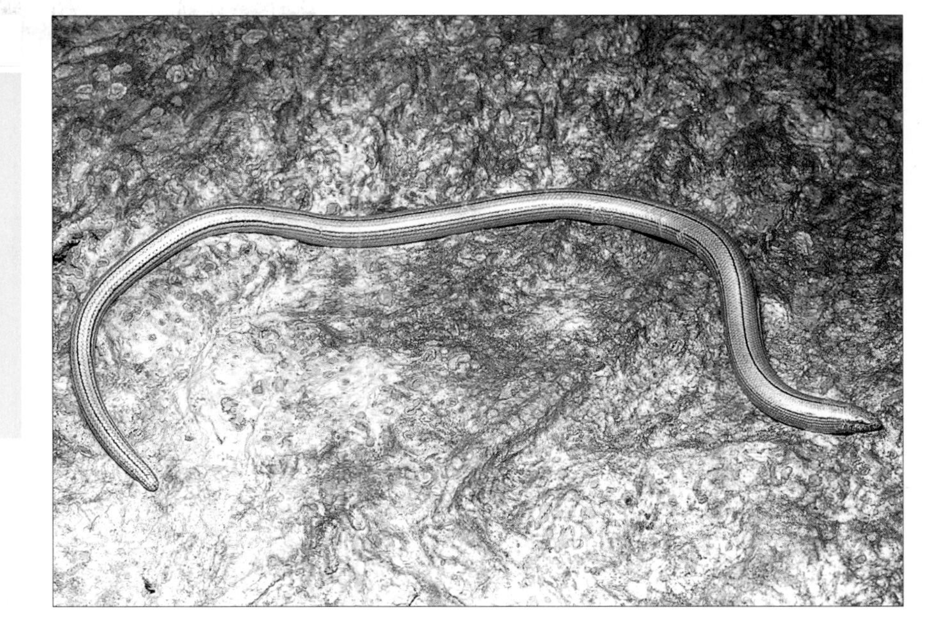

The California legless lizard spends much of its time burowing in search of its soil-living prey, although it sometimes ventures above ground.

Young are born alive

California legless lizards are ovoviviparous: they bear live young. The eggs hatch just before or at the moment of leaving the female's body. The baby lizards are 2½ inches (6.5 cm) long, of which nearly one-third is tail. They reach sexual maturity at the age of 3 years, at a length of 8–10 inches (20–25 cm).

Short legs, big waggle

It is interesting to compare the American legless lizards with the legless lizards of the family Pygopodidae found in Australia. One of them, found over much of that continent and known as Burton's legless lizard, *Lialis burtonis*, is 30 inches (75 cm) long, of which two-thirds is tail and only one-third head and body. This is in striking contrast to the proportions of a snake, in which the tail is usually about an eighth of the total length. Burton's legless lizard has two tiny flaps—all that remain of its limbs—near the vent. It feeds on other lizards known as skinks (family Scincidae). It does not burrow but worms its way through grass, and it is sometimes locally called a grass snake. This is excusable since, like the California legless lizards, it moves like a snake.

Long-bodied lizards with short legs walk at times with considerable sideways flexing of the body. As the legs evolve to become shorter and the body, including the tail, grows longer and more slender, the reptiles take to crawling rather than walking. With the total loss of the legs, the manner of moving over the ground becomes effectively serpentine. This is an advantage, as among the Australian legless lizards, for moving through dense vegetation. It has its limitations, however, for those like the California legless lizards that live underground, because it restricts them to soft soil or loose sand.

LEMMING

The lemming is a rodent linked in most people's minds with mass suicides, but this long-standing association ought to be put into perspective. The mass migration story is usually told about the Norwegian lemming, which is only one of 10 lemming species, all of which are found in the Northern Hemisphere. They include three species of collared lemmings in Arctic Canada, Siberia and European Russia, two bog lemmings of North America, one wood lemming from Norway to Siberia and four species of true lemmings of northern Europe, Asia and America. All have a stout body some 4–6 inches (10–15 cm) long with 1 inch (2.5 cm) or less of tail. They have dense fur, a blunt muzzle, small eyes and small ears hidden in the fur. This article concentrates on the Norwegian lemming, which also ranges across Sweden, Finland and northwestern Russia.

Safe under snow

The Norwegian lemming lives at 2,500–3,300 feet (760–1,000 m) above sea level, above the tree line. In summer lemmings occupy moist, stony ground partly covered by sedges, willow shrubs and dwarf birch. They make paths through the carpet of lichens and rest in natural hollows or cavities in the vegetation. In fall they move into drier areas, at or about the same level as the summer quarters. In winter they usually live under the snow, protected from cold and from predators, building rounded nests of grass that are sometimes left hanging on twigs when the snow has melted. They make extensive tunnels under the snow. Because their food is lichens, mosses, grasses, twigs, buds and bark, wintry conditions do not interrupt their feeding, and in an ordinary winter they continue to breed.

Several litters in a year

Several litters are produced by each female in the course of a year. The gestation period is 16–23 days. Each litter usually contains around seven young, but there may be as many as 14. The offspring are born in a spherical nest of shredded fibers, moss and lichens that is constructed under cover of a rock or in a burrow.

The lemmings' predators include weasels, ermines, rough-legged buzzards, long-tailed jaegers, ravens and other members of the crow family, as well as snowy and great gray owls. In winter the lemmings are safer under their covering of snow, though owls may still hear their movements and pounce accurately on them. Ermines and weasels are also still present in winter, albeit in fewer numbers, so there is markedly less predation during winter.

Mass migrations

Lemmings, like many small rodents, are subject to population fluctuations from one year to the next. They also share with voles, to which they

A collared lemming, **Dicrostonyx torquatus,** *in one of its winter tunnels. It continues to feed on lichens, twigs, mosses and similar foods throughout the long northern winter.*

LEMMINGS

CLASS **Mammalia**

ORDER **Rodentia**

FAMILY **Muridae and Arvicolinae**

GENUS **True lemmings, *Lemmus* (detailed below); collared lemmings, *Dicrostonyx*; bog lemmings, *Synaptomys*; wood lemming, *Myopus***

SPECIES **4 species of true lemmings, including Norwegian lemming, *Lemmus lemmus*; and brown lemming, *L. sibiricus***

WEIGHT
1½–4 oz. (40–112 g)

LENGTH
Head and body: 4–5¼ in. (10–13.5 cm); tail: ¾–1 in. (1.8–2.6 cm)

DISTINCTIVE FEATURES
Stout body; blunt muzzle; small eyes and ears, hidden in fur; dark gray or brown upperparts; gray or cream underparts

DIET
Grasses, lichens, mosses, buds, flowers, bark and twigs

BREEDING
Age at first breeding: usually 30–50 days (female), around 60 days (male); breeding season: usually March–September; gestation period: 16–23 days; number of young: average 7; breeding interval: up to 3 litters per year

LIFE SPAN
Up to 2 years

HABITAT
Arctic tundra and grassland

DISTRIBUTION
Boreal regions of Northern Hemisphere

STATUS
Populations fluctuate wildly

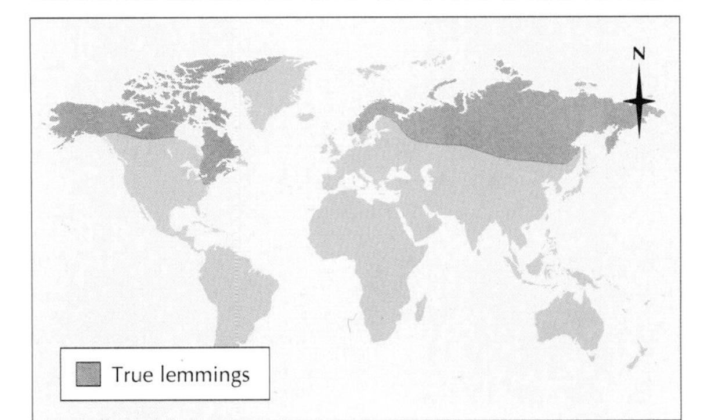

Lemming populations undergo very large fluctuations over a period of years. These are caused by the impact of natural climatic variation on breeding success.

are closely related, more pronounced "boom and bust" population cycles. Numbers build up over a period of years to abnormally high levels; then they plummet before returning again to normal. The interest, so far as lemmings are concerned, lies in the causes of these rises and falls and in what actually happens when their numbers are abnormally high. The scientific explanation in the past has been that in years of abnormal numbers the lemmings migrate down the mountainsides into the fertile valleys in search of food. This is near the truth. The popular stories, aided by artists' impressions based on local hearsay, is of milling columns of lemmings in headlong dash down to the sea, only to drown in what appears at first glance to be a mass suicide. In-depth studies in the latter half of the 20th century began to reveal the truth about the lemmings' behavior, which long ago passed into the English vocabulary to describe reckless.

A matter of climate

The primary factor behind population explosions among lemmings appears to be the vagaries of the climate, with special reference to those years in which a mild spring, late fall and similarly mild winter combine to allow more breeding activity than usual. When the following winter is a little more severe, the lack of enemies during the cold months mean that there is no brake on the mounting increases in numbers.

Panic among the masses

In the years 1960–1961, for example, there was an explosive eruption of lemmings in three waves: one in May, the second in June and a third in

August. These were noticeable mainly in places where obstacles, such as a long lake or the confluence of two rivers, prevented the animals from dispersing evenly. In such instances an accumulation of lemmings is followed by a kind of panic in which they march more randomly. They may go up the mountains as well as down into valleys. They may go to any point of the compass. They may go over glaciers, or swim across rivers or lakes or, as in Norway where the mountains run down to the sea, into the sea. Lemmings swim well, with the body and head well out of water, and their fur is waterproof. If the water is calm, they can cross a river or lake. But in the choppy swell at the mouth of a fjord, or on lakes in windy weather, many drown.

A study in stress

During 1960 there were abundant lemmings but no crash in numbers. This came in 1961, and the explanation is probably supplied by a study of the collared lemming in North America by W. B. Quay. He found that under warm conditions, especially when there was stress or tension, the lemmings suffered an upset of their internal balance. One symptom was abnormal deposits in the blood vessels of the brain. The result was severe exhaustion and finally death.

Tales of mass lemming suicides are nothing more than legends. They arose because of the periodic migrations of lemmings in years when populations are abnormally high.

Another study of Norwegian lemmings has shown that although the animals live solitary lives on the mountaintops and are probably intolerant of each other's company, when they descend into the valleys, they become sociable. For a while, therefore, they feed amicably side by side and share burrows. This may be what happened in 1960. As their numbers rise, so the stress mounts, with a high death rate, as in 1961.

Rains of lemmings

The first published account of the lemming's "suicidal" tendencies was written by Zeigler, a geographer of Strasbourg, in 1532. It was based on information given him in Rome by two bishops from Norway. He retold how in stormy weather lemmings fell from the sky, that their bite was venomous and that they died in thousands when the spring grass began to sprout. In 1718 Joran Norberg, writing of the march of Charles XII's army over the mountains in Norway, said, "People maintain that clouds passing over the mountains leave behind them a vermin called mountain mice or lemmings." Inuits in Arctic America have similar beliefs about rains of lemmings, and their name for one species in Alaska—anticipating modern ideas about UFOs—is the "creature from space."

LEMUR

Lemurs belong to the Primate order, which also includes monkeys and apes. They vary in size from that of a small dog to that of a mouse and live on Madagascar and the Comoros. The more specialized indri, sifaka and aye-aye are treated in separate articles. The remainder can be divided into three very distinct groups. To the first belong the mouse lemurs and dwarf lemurs: small, rarely over 1 foot (30 cm) long, with a long, somewhat bushy tail, short pointed face, and coat colors ranging from red to gray with a white or yellow belly. One species, the fork-crowned lemur, has a pattern of black stripes on its head joining the dorsal stripe to the rings around its eyes. The second group are the true lemurs, 20–40 inches (51–102 cm) long, also with a very long, bushy tail. The black lemur has tufted ears and fringed cheeks. The males and young are black with bright yellow eyes, and the females are pale reddish yellow. The ring-tailed lemur, gray with a black-and-white ringed tail, has a white face with black muzzle and black eye-rings. The ruffed lemur, the largest, is panda-like, black and white or sometimes red, black and white. The gentle lemur is brown with white eye-rings and short snout.

The sportive or weasel lemur stands alone to form the third group. It is about the size of the largest of the dwarf lemurs, with large eyes and ears and a short face. It leaps from one upright stem to another on its long hind legs.

Many lemurs are found almost everywhere on Madagascar, mostly in the forest but also in the drier scrublands. The ring-tailed lemur lives, however, in the southwest of the island, the ruffed lemur in the northeast. The rarest and most localized species are the hairy-eared dwarf lemur, of which only a few specimens are known, thought to be from the eastern forests, and the broad-nosed gentle lemur, equally rare and localized in just a few places in the east.

Sun, sound and scent

The larger lemurs are active mostly at dawn and dusk, although they can be seen on overcast days. Within those limitations, ring-tailed lemurs are more diurnal (day-active), whereas ruffed lemurs are nocturnal. All lemurs like to sun themselves, sitting upright with their arms spread as if meditating. The larger lemurs, especially the black and ring-tailed lemurs, live in social groups of a dozen or so, often with more males than females. The males have a distinct pecking order, but all females are dominant to any male. The animals make grunts to keep the group together, and the ring-tailed lemur makes meowlike sounds, hence the scientific name: *Lemur catta*. The ruffed lemur utters a series of intense roars, followed by loud clucks.

The ring-tailed and gentle lemurs have scent glands on the shoulders and forearms, the latter being provided with spurs, used in marking territories by touching branches with them. The shoulder glands are found only in males. They draw their tails across the glands and then flick them, dispersing the scent into the air. The black lemur rubs its forearms on branches, although it has no glands there, while most other species of lemurs mark with urine or feces. It is believed that these methods of marking are connected with the social ranking of the males.

Ring-tailed lemurs (juvenile, above) are the most sociable of all lemurs. They live in family groups and communicate with grunts and meows.

A male ruffed lemur, Varecia variegata, *in a fig tree. Lemurs feed mainly on ripe fruits and fresh leaves.*

LEMURS

CLASS **Mammalia**

ORDER **Primates**

FAMILY **Cheirogaleidae, Lemuridae and Megaladapidae**

GENUS ***Microcebus, Cheirogaleus, Phaner, Allocebus, Hapalemur, Lemur, Varecia*** **and** ***Eulemur***

SPECIES **At least 36 species, including ring-tailed lemur, *Lemur catta* (detailed below)**

WEIGHT
5–7¾ lb. (2.3–3.5 kg)

LENGTH
Head and body: 15–18 in. (38–46 cm); tail: 22–24½ in. (56–62 cm)

DISTINCTIVE FEATURES
Slender body; very long, bushy tail; black-and-white facial mask and rings around tail; grayish upperparts; white underparts; bright orange eyes

DIET
Mainly fruits, leaves, flowers and bark; occasionally bird eggs

BREEDING
Age at first breeding: 30 months (male), 20 months (female); breeding season: April–July; gestation period: about 135 days; number of young: 1 or 2; breeding interval: 1–2 years

LIFE SPAN
Up to 20 years in captivity

HABITAT
Scrub woodland and riverside forest

DISTRIBUTION
Southwestern Madagascar

STATUS
Endangered

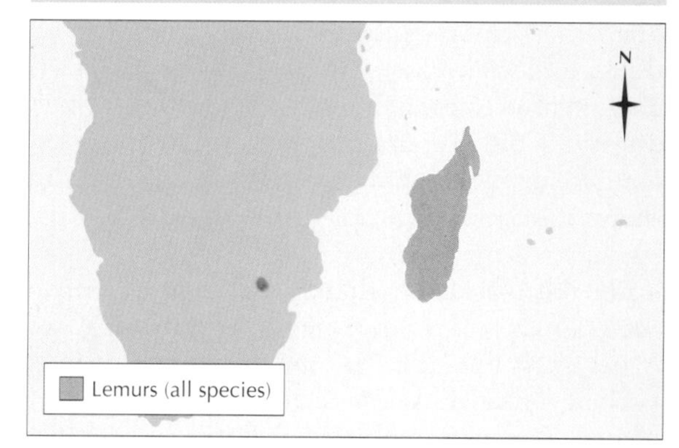

Summer sleep

The mouse lemurs and dwarf lemurs, as well as the sportive lemur, are nocturnal, sleeping most of the day in nests built among foliage or in holes in trees. The two dwarf lemurs are unusual in having a less efficient temperature regulation system than most mammals. The larger of them has short phases of torpor lasting 2 or 3 days, while in the smaller species the torpor may last a week or more. During this period the animal's body heat falls near to that of the ambient temperature. In eastern Madagascar it probably spends most of the dry season in this condition, which is known as estivation. During its torpid period this dwarf lemur survives on fat stored mostly in its tail.

Mainly vegetarian

The sportive lemur is entirely vegetarian, feeding on leaves, fruits, bark and buds. The dwarf lemurs and mouse lemurs eat fruits, berries and insects, but probably not leaves or bark. The larger lemurs concentrate more on leaves and fruits, and some take insects, too, along with bird eggs.

Different kinds of infancy

Lemurs are sexually inactive for most of the year. Mouse lemurs come into season every 35 days or so between September and January. Larger species are in season from April to June, sportive lemurs from May to July; normally each cycle lasts only a month. Mouse lemurs have a gestation of 2 months and have two or three young; other lemurs have a gestation of 4½ months and have only a single young. The dwarf and mouse lemurs bear their young naked and undeveloped, and the mother puts them into a nest made in the foliage. The young ruffed lemur is also born at an early stage of development, and the mother builds it a nest in the fork of a tree, which she lines with fur from her flanks. In other lemurs the young are born fully furred and able to move around. The young black lemur clings to its mother's belly; so does the young ring-tail for a few days, after which it is carried on its mother's back. The young sportive lemur is well developed and moves around on its own near its mother. The mother, when leaping to another tree, carries it in her mouth. The mouse lemur has the fastest development of any primate. It is independent of its mother after 4 months and fully mature at 7 to 8 months. Even so, this remains a reasonably long period of maternal care. Extended maternal care is a feature of mammals, but is most obvious in the primates.

Air raids

Madagascar is relatively free of natural predators. The fossa, a relative of the mongooses and civets, preys on lemurs. The main danger, however, comes from the air. Eagles glide silently through the treetops and may round a corner and snatch up a small lemur at a moment's notice. Eagles are mobbed by ring-tailed and black lemurs, which make a series of grunts repeated every 2–5 seconds. They work up to a crescendo, all in unison. To escape, the black lemur jumps up to 27 feet (8.2 m); dwarf lemurs hide in the foliage, and sportive lemurs leap from trunk to trunk.

A primate by any other name

The lemurs are classed among the prosimians or "lower" primates, along with bushbabies, lorises, pottos and tarsiers. This term distinguishes them from the monkeys, apes and humans, sometimes termed the anthropoid or "higher" primates. The prosimians lack the extreme specializations in binocular vision, high intelligence and manual dexterity that mark the preeminence of anthropoids. Also, in higher primates, the increase in the sense of sight has been linked with a loss in the sense of smell, whereas lemurs have retained a better sense of smell. The apparently primitive nature of lemurs stems from the fact that they evolved in isolation from other primates (and most large predators, too) over a period of 50 million years or more, having probably floated from the African mainland on rafts of vegetation.

Though evolutionarily far removed from the anthropoids, lemurs have nevertheless undergone striking specializations all of their own. They have, for example, a unique and as yet unexplained structure of the ear region of the skull. They have nails on all digits except the "index" toe, which has a sharp claw used for grooming. Their lower incisors and canine teeth are slanted forward and flattened like the teeth of a comb, and this, it seems, is exactly what they are used for: combing their fur. They also have a long, horny filament, the sublingua, underneath the tongue, which is used to scrape the accumulated dirt out of the "comb."

A crowned lemur,* Eulemur coronatus. *The lemurs are a specialized group of primates with a range of unique physical features.

Index

Page numbers in *italics* refer to picture captions.
Index entries in **bold** refer to guidepost or biome and habitat articles.

Page numbers in *italics* refer to picture captions. Index entries in **bold** refer to guidepost or biome and habitat articles.